饒宗頤講座
Jao Tsung-I Lecture

香港大學
饒宗頤學術館
Jao Tsung-I Petite Ecole
The University of Hong Kong

證² ＋ 證³ ＝ 證⁵ ≡ 證 ＝ 一
（二重證據法加三重證據法等於五重
證據法當且僅當終應歸一的證據）
——再論中國古代學術證據法

EVIDENCE² + EVIDENCE³ = EVIDENCE⁵ ≡ EVIDENCE = ONE
(DOUBLE EVIDENCE PLUS TRIPLE EVIDENCE EQUALS QUINTUPLE EVIDENCE IF AND ONLY IF EVIDENCE IS UNITARY)
– FURTHER REMARKS ON THE EVIDENTIAL METHOD FOR SCHOLARSHIP ON ANCIENT CHINA

夏含夷
Edward L. Shaughnessy

香港大學饒宗頤學術館
饒宗頤講座系列 (之三)
Jao Tsung-I Petite Ecole, The University of Hong Kong
Jao Tsung-I Lecture in Chinese Culture No. 3

饒宗頤教授向夏含夷教授贈送其墨寶
Prof. Jao Tsung-i presenting his calligraphy to Prof. Edward L. Shaughnessy.

夏含夷教授與饒宗頤教授
Prof. Edward L. Shaughnessy and Prof. Jao Tsung-i

夏含夷教授演講中
Prof. Edward L. Shaughnessy in the Lecture

夏含夷教授回答現場聽眾提問
Prof. Edward L. Shaughnessy answering questions from the audience.

夏含夷教授與香港大學饒宗頤學術館館長李焯芬教授（右）、學術部主任鄭煒明博士（左）
Prof. Edward L. Shaughnessy and Prof. Lee Chack-Fan, Director of Jao Tsung-I Petite Ecole, The University of Hong Kong (Right), Dr. Cheng Wai-Ming, Head, Research Division, Jao Tsung-I Petite Ecole, The University of Hong Kong (Left).

夏含夷教授與香港大學饒宗頤學術館學術部工作人員
Prof. Edward L. Shaughnessy and Research Division, Jao Tsung-I Petite Ecole, The University of Hong Kong

目錄 Content

香港大學饒宗頤講座簡介
About HKU Jao Tsung-I Lecture
in Chinese Culture 2

內容提要
Abstract 4

證^2 ＋ 證^3 ＝ 證^5 ≡ 證 ＝ 一
（二重證據法加三重證據法等於五重證據法當且僅當終應歸一的證據）
——再論中國古代學術證據法 10
Evidence^2 + Evidence^3 = Evidence^5 ≡ Evidence = One
(Double Evidence Plus Triple Evidence
Equals Quintuple Evidence If and Only
If Evidence Is Unitary)
- Further Remarks on the Evidential Method
 or Scholarship on Ancient China 40

夏含夷教授簡介
About Prof. Edward L. Shaughnessy 74

香港大學饒宗頤講座簡介

　　香港大學饒宗頤講座成立於2012年3月10日，由香港大學饒宗頤學術館主辦，饒學研究基金贊助，以國際知名漢學家饒宗頤命名，講座每年舉辦一次，旨在通過邀請在中華文化研究領域具傑出成就之名家學者來港舉辦高端學術講座，促進海內外的學術文化交流與互動。每位主講的講稿和其他相關文獻編輯成冊，即為「饒宗頤講座系列」，由香港大學饒宗頤學術館出版。

　　饒宗頤教授，字伯濂，又字選堂，號固庵。1917年生於廣東潮州，自學成家，長期致力學術研究，至今有著作約七十餘種，論文近千篇。饒教授是蜚聲國際的百科全書式大學者，素有國學大師之稱，近年更有國寶之譽。他在歷史、文學、語言文字、宗教、哲學、藝術、中外文化關係等人文科學領域中，皆有卓越的成就和突出的貢獻，備受海內外同儕和後輩尊崇。饒教授又是當代最著名的中國傳統文學巨匠，古體、律、絕，無一不精，尤擅填詞，又騷、賦、駢、散，無一不曉，可謂卓立獨行於當代中國文壇，別樹一幟。饒教授更是一位傑出的藝術家，在書法、山水、人物畫的創作上，承先啟後，自成一家，甚至在音樂，特別是古琴，也素有涉獵和創造。饒教授可謂文、藝、學三者兼備，堪稱「一身而兼三絕」，在香港以至於當代的中國，實是百年難得一遇的巨擘。

曾任中山大學廣東通志館專任纂修、研究員；無錫國學專修學校教授；香港大學中文系講師、高級講師、教授；新加坡國立大學中文系首任講座教授、系主任；美國耶魯大學研究院客座教授、臺灣中央研究院歷史語言研究所研究教授；香港中文大學中文系講座教授、系主任；法國高等研究院宗教學部客座教授；日本京都大學文學部及人文科學研究客座等職。現任香港中文大學中文系榮休講座教授、中國文化研究所及藝術系偉倫講座教授；亦是北京大學、浙江大學、南京大學、復旦大學、中山大學、廈門大學、山東大學等著名學府的名譽教授。

曾先後獲得海內外學界和藝壇多個重要獎項與殊榮。其中包括多個博士學位：1982年獲香港大學榮譽文學博士學位，1993年獲法國索邦高等研究院（即巴黎高等實用研究院）授予建院125年以來第一位人文科學榮譽國家博士學位，1995年至今先後獲香港七所大學頒授榮譽博士學位。2006年獲日本創價大學名譽博士學位，2011年獲澳洲塔斯曼尼大學榮譽文學學位。2014年獲山東大學名譽博士學位。更曾獲法國法蘭西學院頒授儒林漢學特賞(1962)、巴黎亞洲學會授予榮譽會員榮銜(1980)、法國文化部授予文化藝術騎士勳章(1993)、香港藝術發展局授予第一屆視藝成就獎(1997)、香港特別行政區政府授予最高榮譽之大紫荊勳章(2000)、中國國家

文物局等聯合主辦之敦煌藏經洞發現百年紀念大會授予敦煌文物保護、研究特別貢獻獎(2000)、(俄羅斯)國際歐亞科學院院士(2001)、中央文史研究館館員(2009)、香港藝術發展局的終身成就獎(2009)、中國藝術研究院中華藝文終身成就獎(2011)、西泠印社社長(2011)、寧波天一閣博物館名譽館長 (2013)等。2011年10月,南京紫金山天文臺更將國際編號為10017的小行星命名為「饒宗頤星」,以肯定饒宗頤教授的卓越成就,褒獎他對中國文化事業乃至對人類文化發展作出的貢獻。2012年12月,饒宗頤教授當選為法蘭西學院美文與銘文學院外籍院士。2014年,獲頒香港大學建校以來首個「桂冠學人」頭銜,是為該校最高學術榮譽。

香港大學饒宗頤學術館創立於2003年,亦即是以饒教授命名,更由饒教授捐贈畢生藏書數萬冊及個人書畫作品而建。香港大學饒宗頤學術館秉承饒宗頤教授立足中華傳統、兼具國際視野和普世關懷之文化創新精神,設立此饒宗頤講座,並以饒教授命名,以志本源。

About Jao Tsung-I Lecture in Chinese Culture, HKU

Established on March 10, 2013, the Jao Tsung-I Lecture in Chinese Culture is organized by Jao Tsung- I Petite Ecole, the University of Hong Kong and funded by Jao Studies Foundation, in honor of Prof. Jao Tsung-I, the world-renowned sinologist. The Lecture is held once per year, inviting distinguished scholar in the field of Chinese culture, to enhance cultural exchange and academic excellence. The speech note of each guest shall be edited and published as independent booklet, forming the *Jao Tsung-I Lecture in Chinese Culture* Series by Jao Tsung- I Petite Ecole, the University of Hong Kong.

Jao Tsung-i, alias Bolian, Xuantang, or Gu'an, was born in Chaozhou, Guangdong in 1917. As a great self-taught academician dedicating his whole life to study, he has published over 70 books and about 1000 papers. Prof. Jao is a world renowned versatile scholar, master of sinology and "living treasure" in China, who has eminent achievements in the research of history, literature, palaeography, philosophy, art, and Sino-foreign relationship. He is also a contemporary traditional Chinese literary master, excelling at Classical Chinese and every types of poetic writing. Furthermore, Prof. Jao is an exceptional artist of painting and calligraphy, learning from previous masters and creating his own style. Regarding the area of music, he also dabbles in Guqin and the composing of musical pieces. Prof. Jao is

well recognized for his great accomplishments in literary creation, artistic, and academic fields, and he is worthy of the praise of being a "giant" of those domains in Hong Kong and even in contemporary China.

Prof. Jao was appointed as Editor and later Researcher of Centre of Local History at Sun Yat-sen University of Guangdong Province; Professor of Wuxi Institute of Chinese Studies, Lecturer, Senior Lecturer and Professor of Chinese Department of the University of Hong Kong; first Chair Professor and Head of Department of Chinese Studies, National University of Singapore; Visiting Professor of Yale University; Research Professor of Institute of History and Philology, Academia Sinica, Taipei; Chair Professor and Head of Department of Chinese Language and Literature, the Chinese University of Hong Kong; Visiting Professor of Ecole Pratique des Hautes Etudes; Visiting Scholar of Kyoto University. He is currently Emeritus Professor of Department of Chinese Language and Literature, Wai Lun Chair Professor of the Institute of Chinese Studies and Department of Fine Arts, The Chinese University of Hong Kong; and is Honorary Professor of various noted universities in China.

Prof. Jao has received many important honors and awards in the cultural and academic fields, which include but not limited to: Various honorary doctorate degrees by HKU (1982) and other seven universities of Hong Kong, the École Pratique des Hautes Études in France (1993), Soka University, Japan (2006), the University of Tasmania in Australia (2011), Shandong University (2014), etc. Awarded Prix Stanislas Julien awarded by the Academie des Inscriptions et Belles-lettres in France (1962), Honorary Member of the Societe Asiatique in Paris (1980), Officier de L'ordre des Arts et des Lettres by the Minister of Culture in France (1993), Grand Bauhinia Medal by the Hong Kong SAR Government (2000), Academician of the International Academy of Sciences for Europe and Asia (2001), Member of the Central Research Institute of Culture and History in China (2009), Lifetime Achievement Award by Hong Kong Arts Development Council (2009) and China Arts Awards Lifetime Achievement Award by the Chinese National Academy of Arts (2011), President of the Xiling Society of Seal Art (2011), Honorary President of Ningbo Tianyige Museum (2013). In 2011, the International Astronomical Union approved the denomination of the minor planet, numbered 10017, discovered by the Purple Mountain Observatory in Nanjing as "Jao Tsung-i", in order to tribute to the marvelous accomplishment of Prof. Jao and his great contribution to the cultural development of China and the world. In December, 2012, Prof. Jao was elected Associé étranger, l'Académie des Inscriptions et Belles-Lettres, l'Institut de France. In 2014, he was appointed the first University Laureate of the University of Hong Kong, the highest academic honour can be offered by the University.

Jao Tsung-I Petite Ecole of HKU was established in 2003 in honour of Prof. Jao and his donation of his dozen-thousands-volume book collection to the University. The department, therefore inherited his innovative insight combined with affection for traditional Chinese culture, international perspective and the universal value, set up this particular Lecture in his name.

內容提要

　　王國維先生所提倡的「二重證據法」是整個二十世紀中國古代文化史上最有名的研究方法。然而，1982年饒宗頤先生又提出了「三重證據法」，就是在王國維「紙上之材料」和「地下之材料」以外，再加上物質文化證據。在2003年，饒先生又談了學術方法，又增加了兩種「間接史料」，即「民族學資料」和「異邦古史資料」。甚麼叫做證據，要怎樣利用證據，要怎樣判斷證據的輕重都是歷史學家的先決問題。王國維和饒宗頤的觀點可以算是現代史學的主流，在我合編的《劍橋中國古代史》裏，我就採取了這個觀點作為該書的基本構造。雖然如此，該書出版了以後，我們發現不是所有的學者都接受這一觀點。有兩種書評發表，從兩個不同的立場提出尖銳的批評，一個說這個做法輕視了中國傳統文獻，一個說太重視傳統文獻的歷史性，因此是「不科學的、不客觀的」。在這個演講，我先對二十世紀的史學觀點做一個簡單的總覽，然後比較詳細地討論《劍橋中國古代史》的編輯工作和讀者反應。

證² + 證³ = 證⁵ ≡ 證 = 一

Abstract

Throughout the twentieth century, the "Double Evidence Method" advocated by Wang Guowei was the most important research method for the cultural history of ancient China. However, in 1982, Jao Tsung-i proposed a "Triple Evidence Method," adding material culture to the "paper sources" and "underground sources" of Wang Guowei. In 2003, Jao again discussed scholarly methods, and added two more "indirect" types of evidence: "anthropological sources" and "ancient historical sources of other countries." What constitutes evidence, how to use evidence, and how to decide the weight to give to evidence are the first problems encountered by historians. The views of Wang Guowei and Jao Tsung-i can be considered as the mainstream of contemporary history, and in *The Cambridge History of Ancient China*, of which I was co-editor, we adopted this methodology as the basic structure of the book. Nevertheless, after the book was published, we discovered that not all scholars accept this viewpoint. Two book reviews were published raising pointed criticisms from two different standpoints: one said that we had under-emphasized traditional Chinese literature, while the other said that we had over-emphasized the historicity of traditional Chinese literature, and because of this our results were "unscientific" and "non-objective." In this lecture, I will first provide a brief overview of twentieth century viewpoints regarding historical research methods, and then will give a more detailed discussion of the editorial work on and readers' responses to *The Cambridge History of Ancient China*.

證² ＋ 證³ ＝ 證⁵ ≡ 證 ＝ 一
（二重證據法加三重證據法等於五重證據法 當且僅當終應歸一的證據）

——再論中國古代學術證據法

　　饒宗頤先生對中國傳統文化與學術領域成就卓著，貢獻特多，從甲骨學與古文字學到簡帛學與上古文獻學到考古學與上古史到文化交流史到敦煌學到歷史學與潮學到宗教到藝術到古典文學，幾乎沒有一個重要問題沒有做過精彩研究。饒先生的學術特點是他的眼光非常廣博，能夠跨越和融合許多專科。根據這種多學科交叉的學術方法，他的研究成果真的遍天下，如果選出這個領域的著作代表他的成果，就對不起那個領域。雖然如此，可能最有超學科意義的貢獻原來不是一篇專門研究，而是饒先生在1982年召開的「香港夏文化討論會」開幕式上的發言。饒先生談了學術方法，說：

　　　　我想借此機會説一説研究夏文化的材料和方法的問題。現在大家都把注意力集中在田野考古中探索夏文化的遺存，這無疑是十分重要的。夏文化的研究能否出現決定性的突破，有賴於這方面的努力。

但是就夏文化的整體而言，地下遺存畢竟有它的局限性，而且遺存也不一定有文字標誌足以表明文化的內涵；所以，我們還得把考古遺存同傳世文獻結合起來進行考察和研究。儘管古籍中關於夏文化的材料不多，但是許多零星的記載，卻往往透露着夏代社會的資訊，有待我們進一步去發掘。

值得特別提出的是甲骨文，在甲骨文中有許多關於商代先公先王記載，在時間上應該屬於夏代的範疇，可看作是商人對於夏代情況的實錄，比起一般傳世文獻來要可靠和重要得多。我們必須而且可以從甲骨文中揭示夏代文化的某些內容。這是探索夏代文化的一項有意義的工作。

總之，我認為探索夏文化，必須將田野考古、文獻記載和甲骨文的研究三方面結合起來，即用「三重證據法」（比王國維的「二重證據法」多了一種甲骨文）進行研究，互相抉發和證明。倘能在這方面做出成績，那麼，我們對於夏代情況的瞭解，將會更加具體而全面。那時來討論夏文化的有關問題，就可說是「適時」了。我們期待着這一天早日到來。[1]

饒先生在這個發言中提出了「三重證據法」，不僅僅是

---

[1] 曾憲通：〈選堂先生三重證據法淺析〉，《華學》九、十合輯（上海：上海古籍出版社，2008年），第2頁所引。

對夏文化有用處,而是對所有學術問題一個基本原則。像饒先生在括號中所提示那樣,「三重證據法」顯然是針對王國維(1877-1927)先生的「二重證據法」而提出來的。王國維的「二重證據法」可以說是整個二十世紀中國古代文化史上最有名的研究方法,在中國國內幾乎沒有學者不贊成。王先生也是在做講話時提出了這個方法論。在1925年,王先生在清華國學研究院講授《古史新證》時提倡:

> 吾輩生於今日,幸於紙上之材料外,更得地下之材料。由此種材料,我輩固得以據以補正紙上之材料,亦得證明古書之某部分全為實錄,即百家不雅馴之言亦不無表示一面之事實。此二重證據法,惟在今日始得為之,雖古書之未得證明者,不能加以否定,而其已得明者,不能不加以肯定,可斷言也。[2]

王國維是文獻學和文字學家,難怪他所提出的學術方法是限制到有文字資料:「紙上之材料」當然是傳世文獻,「地下之材料」乃是出土文字資料,諸如甲骨文、金文和簡帛,甚至包括敦煌各種不同語言的卷子,即使

---

[2] 王國維:《古史新證》,《王國維文集》第四卷(北京:中國文史出版社,1997年),第2頁。

嚴格來說敦煌寫本也是「紙上之材料」。[3] 饒宗頤先生的學術觀點比王國維的還要廣泛，除了傳世文獻和出土文字資料以外，他還注重物質文化，即他所說的田野考古遺存，作為第三種證據。

饒先生提倡了「三重證據法」二十年以後，又有申述，提出「五重證據法」。在2003年出版的《饒宗頤二十世紀學術文集》上的〈論古史的重建〉一篇文章裏，他先回顧了「三重證據法」的說法：

> 余所以提倡三重史料，較王靜安增加一種者，因文物的器物本身，與文物之文字紀錄，宜分別處理，而出土物品之文字紀錄，其為直接史料，價值更高，尤應強調它的重要性。

除了這樣三種「直接史料」以外，饒先生還提倡今後要考慮兩種間接史料，即「民族學資料」和「異邦古史資料」作為互相比勘。這種「五重證據法」和張光直（1931-2001）先生三十五年前所提出「商代之五門」至少表面上有所相似之處。張先生在1980年出版的《商文明》的序論中，說研究商代文明有五種證據，即他所說

---

[3] 因此陳寅恪對二重證據法做了應有的說明，說它的意義有三方面：「一曰取地下之實物與紙上之遺文互相釋證」；「二曰取異族之故書與吾國之舊籍互相補正」；「三曰取外來之觀念，以固有之材料互相參證」；陳寅恪：〈王靜安先生遺書序〉，《金明館叢稿二編》。

的五個「門」:傳世文獻、銅器(包括銘文、器類、紋飾和工藝)、甲骨、考古學以及理論模型。[4]

其實,自從二十世紀初葉以來中國最權威的歷史學家都提倡了相同的「多重證據法」,只是不一定利用了這樣的名稱而已。早在1905年劉師培(1884-1919)先生在他著〈古政原論始論·總敘〉中,主張運用書籍、文字、器物三種證據互證的方法。[5] 同樣,傅斯年(1896-1950)不但提倡要利用這三種史料來做古史研究,他也稱之為「直接材料」。據傅先生説,這些直接材料還不夠:「所以持區區的金文,而不熟讀經傳的人,只能去做刻圖章的匠人。」如果沒有歷史學理論,這只算是「玩古董」。[6] 連王國維在講授《古史新證》時的對手顧頡剛(1893-1980)先生自己也提出了一種多重證據法,説:「我們現在受了時勢的誘導,知道我們可用了考古學的成績作信史的建設,又可用了民俗學的方法作神話和傳説的建設,這愈弄愈糊塗的一筆賬,自今伊始,有漸漸整理清。」[7] 最近,清華大學教授李學勤先

---

[4] Kwang-chih Chang, *Shang Civilization* (New Haven: Yale University Press, 1980), pp.1-65. 另見張光直著,張良仁等譯,《商文明》(瀋陽:遼寧教育出版社,2002年)。

[5] 劉師培:〈古政原始論〉,原刊《國粹學報》1.4(1905年),後收入《劉申叔遺書》第18冊(寧武南氏校本)。

[6] 傅斯年:〈史學方法導論〉,《傅斯年全集》第2冊(長沙:湖南教育出版社,2003年),第310頁。

[7] 顧頡剛:《中國上古史研究講義·自序》(北京:中華書局,1988年),第1-2頁。

生在題作「二重證據法與古史研究」的博客上對這個概念做了一個總結:「幾十年的學術史說明,我們在古史領域中的進步,就是依靠歷史學同考古學的結合,傳世文獻與考古發現的互證。今後對上古時期社會、經濟和思想觀念的探索,還是要沿着這個方向走下去。其實談到其他古代國家和文明的歷史,情形也是一樣的。」他最後說「王國維先生的『二重證據法』是對古史研究的重要貢獻,在八十多年後仍值得我們闡述和發揚」。[8] 總之,無論是「二重證據法」、「三重證據法」抑或「五重證據法」,多重證據法是現代中國史學的主流,沒有多少歷史學家會否認。

雖然說沒有多少歷史學家會否認多重證據法的重要性,但是也不是沒有例外,不然的話,今天的演講就至此結束。今天要講的是我自己的一些史學證據法的經驗,向饒先生和其他同仁請教。十五年以前,我和魯惟一(Michael Loewe)合編了《劍橋中國古代史》(*The Cambridge History of Ancient China*),是一本規模相當大的通史。雖然副題是「從文明的起源到公元前221年」(From the Origins of Civilization to 221 B.C.),但是基本內容只包括中國先秦時代的歷史階段,亦即有文字資料的商周時代(周代包括西周、春秋和戰國各個時

---

[8] 李學勤:〈『二重證據法』與古史研究〉[轉貼 2009-12-6, 11:20:01];載於http://tongzi.blog.hexun.com.tw/41606555_d.html.

代)。我和魯惟一開始做計劃的時候,第一個基本原則是每一個時代都應該有兩章,一章針對它的物質文化作闡述、一章針對它的文字資料。因此,商代就有貝格立(Robert W. Bagley)的〈商代考古〉和吉德煒(David N. Keightley)的〈商:中國第一歷史朝代〉,西周時代有我自己做的〈西周史〉和羅森(Jessica Rawson)的〈西周考古〉,春秋時代有羅泰(Lothar von Falkenhausen)的〈青銅時代的衰落:公元前770-481年的物質文化與社會演變〉和許倬雲(Cho-yun Hsu)的〈春秋時代〉,戰國時代就有陸威儀(Mark Edward Lewis)的〈戰國:政治史〉和巫鴻(Wu Hung)的〈戰國時代的藝術與建築〉,都是兩兩互相對立的。在魯惟一和我合著〈序論〉中,我們特別強調了這種「多重證據法」不僅僅是二十世紀中國史學的主流,並且也是西方漢學的傳統做法。[9]

該書出版了以後,幾乎第一篇書評是在中國發表的,即上海大學歷史學系教授謝維揚所作〈《劍橋中國上古史》讀後——誰識廬山真面目〉[10]。在該書評的開

---

9 Michael Loewe and Edward L. Shaughnessy, *The Cambridge History of Ancient China: From the Origins of Civilization to 221 B.C.* (New York: Cambridge University Press, 1999),特別見第13-14頁;另見魯惟一、夏含夷著;夏含夷譯:〈西方漢學的古史研究:《劍橋中國古代史》序言〉,《中華文史論叢》86(2007年),第11-12頁。

10 謝維揚:〈《劍橋中國上古史》讀後——誰識廬山真面目〉,《文匯報》2001年4月7日。

頭,謝教授宣布了此書不是中國人做的:

> 本書總體基本上沒有認真理會近十多年來古史界,特別是國內先秦史學界和考古學界在探討上述重大、前沿問題上所做的大量工作(從這個意義上說,本書在總體上同國內學者的工作有不小的隔膜)。

這一點批評相當難理解;《劍橋中國古代史》包括長達80頁的「文獻目錄」,大多數都是中國學者的論著。說有某種程度的「隔膜」當然可以,但是說此書的作者「沒有認真理會」中國學者的工作似乎說得過份很多。其實,謝維揚教授還有更激烈的批評,說此書的寫法是「挑戰性的」:

> 讀者們會強烈地感到,對於中國學者來說,本書的寫法在許多方面是挑戰性的。例如,本書對文獻關於商以前中國古代歷史的記載基本上持否定的態度(本書第四章的標題就是:〈商:中國第一個歷史王朝〉,因此儘管自司馬遷以來在中國傳統古史系統裏便有夏朝確定的地位,儘管中國學術界整體上認為夏史問題並非虛妄,以對夏史的研究投入了大量的精力,但在《劍橋中國上古史》中有整整四百多年的夏朝歷史還是被無所顧忌地取消了(唯張光

直先生在所撰〈歷史時期前夜的中國〉一章中有總字數僅1000多字的「夏朝的問題」一節,但張的觀點顯然不能代表本書總體的觀點)。

今天我不打算談談夏朝的歷史性這一重大問題。《劍橋中國古代史》從商朝開始,是因為我們對「歷史」這個概念採取了西方傳統的理解,即使可能是狹義,就是說有文字紀錄才進入了歷史時代;沒有文字紀錄仍然算是「先史時代」。《劍橋中國古代史》是劍橋大學出版社出版的,當然要利用西方傳統史學概念。謝維揚完全可以採取另外一個史學概念,但是僅依這一點全盤地否定《劍橋中國古代史》的成果似乎又是過份的,其實也可以說是「挑戰性的」。

謝維揚的這樣批評是有它的理論根據。謝教授說《劍橋中國古代史》所採取的「多重證據法」(他把它稱作「二元法」)輕視了中國傳統文獻。

如果我們在方法上具有關於古代文獻情況的上述新認識,對於《劍橋中國上古史》如此自信和輕易地否定的許多重大的古代史實,就不會自信到在足夠的反證出現之前便輕率地完全拋卻文獻而自起爐灶。《劍橋中國上古史》寫法的一大特點是「二元法」,即將物質證據與文獻證據平行對待,暗含對中國早期文獻總體不信任之意識（正因為如此，其對於王國維的「二重證據法」也不滿意）。

謝維揚這個書評發表了稍後，他在一個題作「古書成書情況與古史史料學問題」的文章中又討論了《劍橋中國古代史》歷史史料觀：

不久前問世的《劍橋中國上古史》（*The Cambridge History of Ancient China*）也許就是一個最近的例子。作為西方漢學界的一部代表性力作，其對於古史史料學問題照例是有敏感意識的（這從它對資料問題的大量論述中可以看出），而它引人注目的一點是在撰寫結構上採用了並不多見的將地下出土資料與文獻資料作「二元」處理的辦法。這給人的印象是：一、在複雜的古史史料學問題面前，它希望避免在每一處關乎古史史料學原則的問題上簡單表態；二、與此同時，它又試圖通過這種「二元」的辦法宣示它的依據地下出土之文物講古史乃古史史料學之正宗的觀念，而文獻的地位則相應變成了「依賴性的」（dependent）。因此其雖沒有正面地將傳世文獻一棍子打死，但依其寫法，文獻作為獨立的古史史料的地位則曖昧了許多，甚至基本上沒有了獨立性，據文獻來講古史先成底氣不足之事（也正因為這樣，儘管文獻中關於夏朝歷史的資料是眾所周知的,但該書卻堅持不寫「夏朝」這一章)。這種古史史料觀也許並不是目前眾多學者所能認同的。[11]

我想我們可以不去管謝教授說我們要「將傳世文獻一棍子打死」;一群老外學者哪兒有資格將傳世文獻一棍子打死?謝教授還說我們所利用的二重證據法一個目的是要將傳世文獻變成依賴性的證據。他說這種二重證據法「不是目前眾多學者所能認同的」,不知道這「眾多學者」都是誰?他們是不是也不能認同劉師培、傅斯年、王國維、饒宗頤和李學勤的史學概念?

---

[11] 謝維揚:〈古書成書情況與古史史料學問題〉,載於吉林大學古籍研究所編《金景芳教授百年誕辰紀念文集》(長春:吉林大學出版社,2002年)。謝氏顯然最重視的問題乃是夏朝的歷史性;在本文後頭他接著說:

> 這說明就中國早期文獻的實際而言,堅持證真方舉證立場的效果是不好的。在古史研究實踐中,證真方舉證立場的影響也值得反思。前文提到《劍橋中國上古史》不列「夏朝」一章,其實《史記·夏本紀》中對夏世系的記載在質量上同〈商本紀〉(引者按:然)沒有根本不同;倒退一百五十年,依《劍橋史》的立場,「商朝」一章也應該不會寫,因為那時不會有殷墟甲骨的「證真」,然而現在沒有人懷疑商史是真實的。因此在〈夏本紀〉的問題上,《劍橋》目前堅持的證真舉證立場實際上反映了其拒絕對中國早期文獻的全面和總體的表現與特徵作完整的思考。其實,依最平實的邏輯推斷,依證真舉證立場排斥〈夏本紀〉很可能會是武斷的。這個事實有很深刻的含義,那就是對於中國早期文獻文本的生成的基本理由應有恰如其分的認識,這一點還是需要我們大力研究的。

將來,如果有像殷墟卜辭的夏代文字證據出現,肯定會有學者重寫《劍橋中國古代史》。我自己深識,根據十五年以來的考古新發現,《劍橋中國古代史》這本書已經應該重寫。然而,關於夏朝的歷史性這一點,我們採取的做法還是按照西方傳統的史學原則,當時文字證據沒有出現之前,恐怕至少這一點不需要重寫。

奇怪的是，有關《劍橋中國古代史》還有與謝維揚的批評完全相反的批評。美國加州大學洛杉磯分校教授（現在任該校文學院院長）史嘉柏（David Schaberg）先生也做了兩篇文章來討論《劍橋中國古代史》，一篇是篇幅很長的書評 "Texts and Artifacts: A Review of the *Cambridge History of Ancient China*"（〈文獻與實物：《劍橋中國古代史》的書評〉），發表在西方權威學術刊物 *Monumenta Serica*（《華裔學志》）上，[12] 另外一篇原來是史教授在華東師範大學做的演講，題作「近十年西方漢學界關於中國歷史的若干爭論問題」。[13] 在他的演講中，史嘉柏就提出了問題：

> 以下，先看芝加哥大學東亞系教授夏含義（引者按：然）先生與普林斯頓大學考古系教授貝格利（引者按：然）先生兩位學者在編寫《劍橋古代中國史》過程中發生的關於歷史方法與考古學方法相對價值的衝突。……辯論題目是甚麼呢？投稿者裏面的考古學家如貝格利、羅泰等人在不同程度上認為要瞭解中國古代史必須擺脫傳世文獻的架子而直接在考古發掘文物之基礎上建立新的歷史解釋，然

---

[12] David Schaberg, "Texts and Artifacts: A Review of the *Cambridge History of Ancient China*," *Monumenta Serica* 49（2001），pp.463-515.

[13] 史嘉柏：〈近十年西方漢學界關於中國歷史的若干爭論問題：2005年10月27日在華東師範大學的學術演講〉，《海外中國學評論》2（2007年），第47-55頁。

後再以研究結果糾正補充舊文獻上之缺點。而投稿者裏面的歷史學家則比較重視文獻的歷史性，以古代史書如《尚書》、《竹書紀年》、《逸周書》、《左傳》等定為出發點，再旁參其他傳世文獻、出土文獻等，即使不能完全使其一致，但最後還可以尋出所有資料之間的相同性。據說在修改過程中雙方互相批判得還相當嚴厲，而最後兩方立場幾乎無動搖。[14]

有關「修改過程」，我自己當作《劍橋中國古代史》的合編者之一，大概是最清楚的。説有「兩方立場」沒有錯，説「雙方互相批判得相當嚴厲」，説「兩方立場幾乎無動搖」恐怕也是説得過分一點。無論如何，《劍橋中國古代史》的作者都同意了該書的二重構造。史嘉柏教授此次演講的結論對這「兩方立場」劃出一個似乎很公平地選擇：

在漢學正在大規模地國際化的關鍵時刻，最令西方學者疑惑的就是一個很基本的問題：要瞭解中國的過去，主要的思考模式要到甚麼程度采自中國文化原有的模式？也就是說對中國古代歷史的瞭解應該是全新的、全科學化的、全客觀的呢？還是應該盡

---

[14] 史嘉柏：〈近十年西方漢學界關於中國歷史的若干爭論問題：2005年10月27日在華東師範大學的學術演講〉，第49-50頁。

量融合歷代中國人本身對過去的一些看法？一定要有所取捨，只是還不知道要採取甚麼，要捨棄甚麼。我認為，進行這種辯論就是歷史學家的任務之一，順着中國文化的全球化，這辯論也應該就是各界漢學者可以共同參加的。[15]

誰能拒絕「全科學化的、全客觀的」學問（據我想，「全新的」學問不一定同樣肯定；我覺得有相當多的「舊」學問仍然值得參考）。然而，倘若深入一點思考，這個選擇並不公平。　問題是，史嘉柏將「全科學化的、全客觀的」的學問置於一邊，與之對立的是「盡量融合歷代中國人本身對過去的一些看法」。難道中國人本身對過去的看法不可能是全科學化的、全客觀的呢，更不用說全新的？其實，這樣的選擇是虛假的。甚麼叫做「歷代中國人本身對過去的一些看法」？按照史嘉柏在他對《劍橋中國古代史》書評的論述，這就是利用傳世文獻作為史料。他說這樣的學問是「是根據信仰的」，也就是說是不科學的，不客觀的。

　　大體來說，歷史學家所利用的文獻都有晚起抑或不明的著作背景。要麼是所敍述的內容發生好幾世紀以後才寫的（諸如《史記》），要麼我們根本

15　史嘉柏：〈近十年西方漢學界關於中國歷史的若干爭論問題：2005年10月27日在華東師範大學的學術演講〉，第47-55頁。

不知道是甚麼時候寫的、是甚麼人寫的。即使是出土文字資料，諸如甲骨文和銅器銘文，基本上也不帶有信息說明其著作條件與用處，更不用說其對相關歷史事件的關係。並且，出土文字資料完全確認了傳世文獻的那些非常罕見的例子（最有名的例子是《史記》所列商代諸王的名單）並不能表明傳世文獻的普遍正確性。如果「傳統歷史」是部分根據傳世文獻的歷史研究，那麼它的出發點要麼是過分不定的、要麼是根據信仰的。歷史學家所應用的歷史框架（包括年代框架）及其細節的信息都引用自很晚或不明的傳世文獻。因此他們對古代的敘述永遠不會像考古學家的構擬那樣確切。

最好的歷史學家利用傳世文獻的時候，以之為來源可疑的文物。他們注重文獻可疑之處、批判對其來源最精銳的解說，而後才作出不過於證據局限的結論。儘管歷史學與考古學的方法不同，但是貝格利對證據的利用、他對證據不足與時間錯誤的文化一同說的否認，及其對公元前第二千年的最底文化交接應該給歷史學家作為一個方法論的典範。……傳世文獻已經失去了它原來製作的許多物質性質。它經過了抄寫、編輯和收集過程——我們也不知道甚麼時候抑或由甚麼人——現在幾乎不可能重構它原來的形式，即使是竹簡上的寫本、口頭

的詔令或者另一種題材的表演。¹⁶

在這篇書評的結論上,史嘉柏就毫無疑問地說出他自己的見解:「我們應該認為傳世文獻不是十分可靠,也不應該當作歷史證據。」¹⁷

謝維揚和史嘉柏的立場可以說是完全對立的,但是兩位學者都針對同一本書指出激烈的批評。難道此書真的壞到這個地步?也許應該說書好到這個地步。無論如何,我想趁「饒宗頤講座」這個機會來談談歷史證據的多重性和單一性。像史嘉柏說那樣,傳世文獻本身就有相當不可靠的性質。確實,有的傳世文獻「我們根本不知道是甚麼時候寫的、是甚麼人寫的」。更嚴重的,所有先秦時代傳世文獻都經過了相當漫長的傳授過程才被寫定於漢代。現在在中國國內有相當普遍的「信古學術潮流」。這是根據近來三四十年出土的戰國秦漢的簡帛文獻,這些出土文獻實在非常重要,沒有人能夠否認。李學勤先生在有名的「走出疑古時代」演講中提倡「學術史一定要重新寫」,說:

> 不管怎麼說吧,我們的想法是現在出土的很多東西

---

¹⁶ David Schaberg, "Texts and Artifacts: A Review of the *Cambridge History of Ancient China*," p.475.

¹⁷ David Schaberg, "Texts and Artifacts: A Review of the *Cambridge History of Ancient China*," p.507.

可以和傳世的古籍相聯繫,像《鶡冠子》,雖然沒有出土,但它和帛書《黃帝書》很像,可以說明《鶡冠子》確實是楚人作的,而且也比較早。像這些例子,可以給我們提供一些定點,可以作出很多的推論。它的趨向是很明顯的,就是和疑古思潮相反。[18]

雖然最近三四十年以來出土簡帛文獻的趨向實在「和疑古思潮相反」,但是這並不意味着傳世文獻與先秦文獻就完全一致。無論是疑古思潮懷疑的文獻諸如《鶡冠子》抑或最可靠的文獻諸如《詩經》,我們現在所看的都經過了漢人的校讎工作,與原貌有一定的距離。我經常給學生提到俞樾(1821-1907)在他所著的《古書疑義舉例‧前言》所說的話,「執今日傳刻之書而以為是古人之真本猶聞人言筍可食,歸而煮其簀也」。這個當然很可笑,但是不僅僅是笑話而已;它的意義很深。其實,李學勤先生在「走出疑古時代」中也體會到這一點:

> 古書的面貌和我們的想像是大不一樣的,這一點我們要有充分的認識。有時候我常常說,我們應該用我們的感受去體會孔安國,或者束皙、荀勖這些人的重大成果。孔安國作隸古定,那時候他對戰國文

---

18 李學勤:《走出疑古時代》(瀋陽:遼寧大學出版社,1994年),第16頁。

字畢竟不大懂，所以弄出很多問題來。[19]

我敢說不但孔安國、束晳和荀勗「弄出很多問題來」，並且劉向和劉歆也弄出很多問題來。有的這些問題已經由歷來訓詁學者指出，但是可能更嚴重的是我們還沒有認識出來的問題。先秦文獻在某種程度上能夠通讀是由於漢人的編輯工作，而這個編輯工作在某種程度上不能不反映漢人的知識背景。因此，說出土簡帛文獻的趨向和疑古思潮相反一點也不錯，但是如果像謝維揚那樣提倡傳世文獻的可靠性恐怕也不一定靠得很住。

傳世文獻如果靠不住，我們是不是要像史嘉柏說那樣，「要瞭解中國古代史必須擺脫傳世文獻的架子而直接在考古發掘文物之基礎上建立新的歷史解釋，然後再以研究結果糾正補充舊文獻上之缺點」？我當然不會否認考古資料對我們現在瞭解中國古代史所起的作用。像李學勤先生在「走出疑古時代」那個演講中差不多第一句話所說，「考古發現對研究歷史作用很大。這一點，恐怕現在所有的人都承認。這點恐怕是一個常識」[20]。然而，考古證據本身不是沒有問題的。在《劍橋中國古代史·序言》裏，討論中國歷史資料的時候，我和魯惟一完全承認了考古資料的價值，但是也指出了這個資料

19　李學勤：《走出疑古時代》，第6頁。
20　李學勤：《走出疑古時代》，第1頁。

的某些缺點:

> 雖然如此,正像文獻資料一樣,考古資料也不無問題和偏見。大陸的考古發掘通常配合基本建設,考古學家能對某一遺址做專門尋找、探索、發掘是極其罕見的。因此,出土的大多數文物的發現都是偶然的,它的地下的存在也是偶然的。中國考古基本上是墓葬考古,所發現的文物大多數是出於墓葬;相反的,沒有放進墓葬的物質資料就沒能保存。地上的文物都受到了種種破壞,無論是大自然的還是人為的。古人選擇何物放進墓葬也有不同的動機。我們可以毫無疑問地說,所發現的文物只能代表古時物質文化極小的一部分。並且,大多數也只能代表古代上層社會之生活。但是,因為我們所有的證據就是如此,有人會覺得這些就是所有的證據。我們應該時時提醒自己,對古代物質文化所不知道的東西比知道的東西要多。還有一點,正如本書幾位作者所指出的那樣,中國考古學也會受到當代政治和文化的影響,其中最重要的莫過於中國學者要弘揚中國文化的偉大性和獨特性的願望。[21]

這些問題是考古學家知道得很清楚,但是某些文獻學家

---

21 Loewe and Shaughnessy, *The Cambridge History of Ancient China*, p. 11; 另見魯惟一、夏含夷著,夏含夷譯,〈西方漢學的古史研究:《劍橋中國古代史》序言〉,第11頁。

像史嘉柏那樣對考古學有相當浪漫態度，把考古資料理想化。倘若我們專門根據考古資料來寫中國古代史，那個歷史恐怕會很貧乏。史嘉柏提出了羅泰作為他的學術模範。羅泰當然是一個非常優秀的學者，但是恐怕連他也不能當作所有的人的模範。幾年前，我給他所著的《孔子時代的中國社會（公元前1000-250）：考古證據》一大作寫了書評，結論謂：

> 在引言和結論裏，羅泰一再刻意迴避傳世文獻，偏信地下出土的考古材料能更客觀地描繪當時的情形。對自己的學科有信心是好的，但也容易弄巧成拙。學科細化所提倡的，好像不是學科間彼此不容的鬥爭，相反，在撥開人類社會重重迷霧的荊棘之路上，唯有學科交叉、綜合才能讓我們抵達最終的目的地。從羅泰對古代墓葬「一根筋式」的探求中，我們確實對那時候人們的死亡瞭解得很多了，但對他們死前的生活卻不甚了了，更不用說對他們的愛情是一無所知。在羅泰的論證裏，詩歌失去了地位，但我們知道詩歌（尤其是《詩經》）在那個時代舉足輕重。雖然我願意瞭解當時的墓葬，但我堅信它們並非打開一切大門的鑰匙。我覺得，在研究古代社會時候，學者們倘若忽視他自己學科以外的資料，也會對所有的資料刨根問底，窮盡闡發。

……

因為羅泰將當時的考古文物與文字資料相互對立，堅持出土文物更有資格見證社會的興衰變遷，必然招致如是批評。在我看來，人們才剛剛意識到孔子時代的社會、歷史實際有多麼的複雜，他們正嘗試着用更為恢弘、交錯的目光回溯既往的歷史，回憶先輩曾經走過的某段歲月，回顧孔子時代。單個學科真的能夠照顧到這種複雜性嗎？我對此表示深刻的懷疑。[22]

傳世文獻如此有問題，考古資料也有問題，難道歷史學就沒有希望，沒有證據可以依靠？這顯然不是。所有的證據，無論是傳世文獻、考古資料、出土文字資料，還是各個學科的理論抑或異邦的對比，儘管都有問題，可是也都有它的價值，都有它的用處。只要我們考慮到它的限制——它本身的偏向，也考慮到我們自己不足之處——也就是說，我們自己的偏見，我們對中國古代歷史應該可以作出某些貢獻。這些貢獻恐怕永遠達不到全新的地步，更不用說全科學化、全客觀的地步。連饒宗頤先生恐怕也達不到這個地步。但是只要像饒先生那樣把我們的眼光多打開，不僅僅局限於某一種證據，反而採取二重、三重，甚至更多重證據法，我們就會像

---

[22] Edward L. Shaughnessy, "(Review of) *Chinese Society in the Age of Confucius（1000-250 BC）: The Archaeological Evidence*," *Journal of Asian Studies* 66.4（2007）: 1132.

饒先生所說那樣作出「適時」的研究成績。饒先生說他「期待着這一天早日到來」。有他作為我們的典範，那一天就更接近了。

問　答

　　宣讀這篇文章以後，在座的諸位教授和同學們提出了不少很好的問題，很多都圍繞了證據用法問題。儘管我同饒宗頤先生一樣提倡利用多重證據法進行中國古代文化史研究，可是問題並不這麼簡單。正如幾位同學提出的那樣，如果兩種證據不協調的時候，更不用說兩種證據完全抵觸的時候，應該怎樣選擇利用一種證據而排除另外一種證據。這樣的問題提得非常恰當，我當時用了十幾分鐘回答，提出了某些實例來談談史學做法，但到最後仍然覺得我沒有作出很妥當的答覆，自己很不滿意。提問題的同學們很客氣，但是到最後恐怕他們也不滿意。現在想趁發表這個小冊子的機會來重新考慮這個關鍵問題，看是不是能夠作出好一點的答覆，滿足他們的要求。

　　做演講的時候，我引用了饒宗頤先生在2003年發表的「論古史的重建」中所提出的「五重證據法」。所謂的五重證據，就是他曾經提出的三重證據，即傳世文獻、出土文字資料和物質文化因素，再加上「民族學資料」和「異邦古史資料」以互相比勘。饒先生自己把這五種證據分成「直接史料」（即前三種）和「間接史料」（後兩種），似乎暗示「間接史料」與「直接史料」互相對比的時候,沒有同等的學術價值。間接史料可以提供比勘,也可以給學者啓發,可是在缺乏其他直接史料的條件之下,不能當作獨立的證據。其實,據饒先生自己說,三種直接史料也不一定都享有同等價值:

> 余所以提倡三重史料,較王靜安增加一種者,因文物的器物本身,與文物之文字紀錄,宜分別處理,而出土物品之文字紀錄,其為直接史料,價值更高,尤應強調它的重要性。

這裏雖然沒有說清楚,但是他的意思好像是說出土資料比傳世文獻更為重要,而出土資料本身要「分別處理」,文字紀錄比出土物品「價值更高」。這樣理解如果不誤,饒先生對多種證據的相互重要性的評價如下:

> 出土文字紀錄
> 出土物品
> 傳世文獻
> 民族學資料和異邦古史資料

我覺得這樣的層次大體上很合理,然而仍然有討論的餘地。下面想介紹一下西方漢學某些權威學者的看法,以便更深入地思考這個問題。

早在考古學證據奪取中國古代文化史證據的權威寶

座之前,西方漢學家高本漢(Bernhard Karlgren; 1889-1978)已經對傳世文獻的相互價值做了評價。他在1946年發表了〈中國古代傳說與禮拜〉("Legends and cults in ancient China"),對其他西方漢學家的學術研究做了總攬,提出了某些批評,特別是有關他們的證據用法。在高本漢看來,中國古代文獻應當分成兩類。第一類「自由文獻」("free texts")都是先秦文獻,諸如《詩經》、《尚書》、《左傳》、《國語》、《戰國策》、《論語》、《孟子》等多少能夠反映當時真實歷史情況的文獻。與此不一樣的是一種「系統化的文獻」("systematizing texts"),諸如《周禮》、《禮記》、《呂氏春秋》、《史記》等,多為漢人編纂的,內容雖然可能包含某些古代思想,但是總的來說表現一種理想化的古代。按照高本漢的總攬,當前西方漢學家的研究成果,諸如葛蘭言(Marcel Granet, 1884-1940)的《中國古代跳舞與傳說》(*Danses et légends de la Chine ancienne, 1926*)和夏德(Friedrich Hirth, 1845-1927)的《中國古代歷史》(*The Ancient History of China, 1908*)都是根據這種「系統化的文獻」寫成的,因此多半都是失敗的。連馬伯樂(Henri Maspero, 1882-1945)的《古代中國》(*La Chine antique, 1927*)也不能完全避免這個方法論的問題。儘管高本漢這樣的證據用法雖然在中國沒有引起多大的注意,但是在以後的西方漢學界影響特別大。[23]

　　到了上個世紀七十年代，有的西方漢學家，比如吉德煒，比高本漢更進一步，認為文字資料可以分成兩個大類型：硬性資料和軟性資料。吉德煒是甲骨文專家，專心地研究了商代歷史。據他來說，所有傳世文獻，無論是高本漢的「自由文獻」還是「系統化的文獻」，都應該算是軟的證據，不可作為中國古代文化史上的獨立證據，出土文字資料才可以是硬的證據。因此，他提倡研究商代歷史要完全依靠甲骨文證據。他這樣很「硬」的方法得到了很多西方學者的讚揚。譬如，在本次演講裏，我曾討論了史嘉柏對《劍橋中國古代史》做的書評。史嘉柏特別提出吉德煒是史學家的典範，認為他的證據用法最為謹慎，其他史學家也應該專門利用「硬」性證據。[24]

　　西方學者當中，還有另外一些學者諸如《劍橋中國古代史》的作者之一貝格理以及本次演講所討論的羅泰教授都認為所有文獻，無論是傳世文獻還是出土文獻，都不如考古學家發掘的物質文化客觀可靠。上面已經對羅泰觀點做了評價，於此似乎可以不用再商量。

---

[23] Bernhard Karlgren, "Legends and cults in ancient China", *Bulletin of the Museum of Far Eastern Antiquity* 18（1946）：199-365。在馬悅然給高本漢做的傳記《高本漢：一個學者的肖像》（*Bernhard Karlgren: Portrait of a Scholar*）裏，他指出這一長篇文章是高本漢唯一沒有翻譯成中文的文章；見 N.G.D. Malmqvist, *Bernhard Karlgren: Portrait of a Scholar* (Bethlehem, Penn.: Lehigh University Press, 2011), p.229.

[24] David Schaberg, "Texts and Artifacts: A Review of the *Cambridge History of Ancient China*", pp.476-477.

我自己的觀點比較模糊，沒有一定的原則和方法。如果一定要採取一個證據層次系統，大概會像饒宗頤先生的那樣排出高低：

> 出土文字紀錄
> 出土物品
> 傳世文獻
> 民族學資料和異邦古史資料

我也很羨慕吉德煒的謹慎史學態度，但是我覺得這樣「謹慎」的態度可能適合商代歷史研究，但是不一定能作其他歷史階段與領域的模範。商代本來沒有多少傳世文獻（也許根本沒有傳世文獻，《尚書》裏〈商書〉各篇可能都是周代以後才寫成，儘管可能保存某些商代資料），歷史學家可以僅以甲骨文為證據，也不失去多少證據。然而，周代文化史（更不用說秦漢時代文化史）的情況迥然不同，恐怕不適合這種研究方法。要研究西周時代歷史，僅以當時甲骨文和金文為史料當然可以，但是正像我在上面批評羅泰那樣，我覺得我這樣做會失去當時最核心的情感。對我來說，僅以出土資料進行周代研究實在過於謹慎。我們要進行周代文化史研究的時候，恐怕還是要採取多重證據法，也就是說要利用出土文字資料、出土物質資料和傳世文獻。

說到這裏，恐怕還沒有給在座的同學們一個妥當的答覆。他們的問題是，兩種證據不協調或甚至矛盾的話，應該憑甚麼條件來作一個選擇？我覺得我們做這樣選擇的時候，不但要說明某一種證據為什麼可靠，還要說明另外一種證據為什麼不可靠，不可靠的證據是在甚麼時候創造的，是為了配合甚麼歷史需要創造的。我利用我比較熟悉的西周年代學來舉一個簡單的例子。在多半的史書諸如《通鑑外紀》、《皇極經世》、《通考》等，傳統說法以為周昭王在位51年。然而，西晉時代出土的《竹書紀年》載有昭王在位僅19年而卒。現在一般歷史學家都同意《竹書紀年》的在位年數確切可靠，也有某些西周銅器銘文似乎可以旁證。然而這51年是怎麼創造的？我曾經推測這個數字很可能來自下列幾個年代學的因素：周穆王在周王朝開始以後一百年即位；周武王既克商後二年而崩；周公攝政七年；成王和康王在位時四十年「刑措不用」。把下面三個因素加在一起（2+7+40=49），再將這個數字從上面100年減去，就得到昭王在位的51年。這似乎表明傳世文獻的這個數字是後人推算出來的，因此不應該視作原始證據。這也並不奇怪，與一般的史學原則一致。這不一定說明出土文字資料（即《竹書紀年》）確實可靠，但是至少比傳世文獻更有史學價值。

然而，下面這個例子說明原則不一定總是如此簡單。上面剛剛提到周武王既克商後二年而崩的傳統說法。這個說法原來在《尚書·金縢》篇有所記載，然

後又在《史記》的〈封禪書〉和〈周本紀〉都有所引用。儘管《逸周書·明堂解》還有一個說法，謂「周公相武王以伐紂，夷定天下。既克紂六年而武王崩」，與〈金縢〉的說法明顯矛盾，但是一般歷史學家以為《尚書》比《逸周書》可靠，因此多採取「武王既克商二年而崩」的說法。雖然如此，前幾年清華大學發表了題為〈周武王有疾周公所自以代王之志〉的一篇戰國竹書文獻，內容和語言與〈金縢〉大同小異，應該視作〈金縢〉的最早寫本。清華本〈金縢〉與傳世本〈金縢〉的一個不同是清華本開頭明確提到「武王既克殷三年」，似乎與傳世本〈金縢〉矛盾。按照出土文字資料優先原則，我們是不是應該採取清華竹書的證據，推定周武王既克商後三年才死，而拋棄傳世本〈金縢〉，更不用說《逸周書·明堂解》的證據？我覺得不一定。雖然古書裏不無旁證，諸如《淮南子·要略訓》謂「武王立三年而崩」，但是由於古代漢語用數字習慣比較模糊，「二年」在某些上下文之間可以等於「三年」（譬如說，某王在位二年而崩即到第三年而崩），也就是說「第三年」可以理解為「兩年以後」。然而，相反的邏輯不行。無論如何，「三年以後」不可等於「二年」。這樣的話，我們可以說明〈金縢〉的「武王既克商後二年而崩」怎麼會變形成為〈周武王有疾周公所自以代王之志〉所載的「武王既克殷三年」，但是不可說明「武王既克殷三年」怎麼會變形成為「武王既克商後二年而崩」。當然,我們可以推測傳世本〈金縢〉在漫長的流傳過程當中是由某某抄手抄錯了,但是這樣的解釋似乎過於偶然。

不知道這個例子能夠說明甚麼原則。也許這只能說明香港大學的同學們所提的問題非常好,或者說非常難。我們可以作出簡單的答覆,但是進行歷史研究的時候一定會遇到種種例外。為了作出一個總結,也許只能再提出本次演講的題目:有多重證據,這些證據都可以使用,但是使用的原則應該是所有的證據都有它的價值。有的證據可以作為直接證明,有的證據可以作為間接旁證,有的證據可以作為反證。只要我們認清證據的創造背景、創造目的,我們都可以利用。

饒宗頤講座
Jao Tsung-I Lecture

Evidence^2 + Evidence^3 = Evidence^5 ≡ Evidence = One
(Double Evidence Plus Triple Evidence Equals Quintuple Evidence If and Only If Evidence Is Unitary)

- Further Remarks on the Evidential Method for Scholarship on Ancient China

    Professor Jao Tsung-i 饒宗頤 is one of the giants in the study of traditional Chinese culture, with contributions ranging from the fields of oracle-bone studies and paleography, to bamboo and silk manuscripts and ancient textual studies, to archaeology and ancient history, to cultural exchange and Dunhuang Studies, to historiography and the study of his native Chaozhou 潮州, and from art history to poetics; there is hardly an important topic in Chinese culture for which he has not produced insightful studies. The most characteristic feature of his scholarship is its extraordinarily broad vision, the way he can transcend and synthesize different scholarly disciplines. Through this multi-disciplinary method, his research results truly cover the world. It would be impossible to choose any single work of his to represent the entirety of his scholarship. Nevertheless, for the purposes of today's lecture, perhaps I

---

*   This lecture was originally written and presented in Chinese, the present translation being by the author. I have tried not to diverge from the original structure, insofar as normal English usage permits.

can take inspiration from one of his most cross-disciplinary statements, one which, however, was never formally published, even though it is well known to scholars in many different fields. This was the opening statement he made at the "Hong Kong Conference on the Xia Dynasty" in 1982. Professor Jao addressed the topic of historiography, saying:

> *I would like to take advantage of this opportunity to talk about the question of the materials and methods for studying Xia Culture. Nowadays everybody focuses on the remains of Xia culture unearthed by archaeology, and there is no doubt that these are very important. If research on Xia Culture is to produce any breakthroughs, it will surely rely on efforts in this regard.*
>
> *However, in terms of Xia Culture as a whole, underground remains ultimately have their limitations. What is more, these remains will not necessarily have any written expression sufficient to express the nature of the culture. Therefore, we have to put archaeological evidence together with transmitted literature in order to carry out our research. Even though ancient texts do not contain many sources concerning Xia Culture, still a number of miscellaneous records occasionally reveal something about the Xia society, and are there for us to discover.*
>
> *Especially worthy of mention are oracle-bone inscriptions, in which there are many records of past ancestors and past kings of the Shang dynasty, who in terms of time would belong to the period of the Xia dynasty. These can be seen as true records*

of the Shang people concerning the Xia, and are therefore much more important than ordinary transmitted literature. From these oracle-bone inscriptions, we must—and can—reveal various aspects of Xia Culture. This is a very meaningful project in exploring Xia Culture.

In short, I believe that to explore Xia Culture, we must join archaeological evidence, literary records and oracle-bone inscriptions; that is to say, to use a "Triple-Evidence Method" (adding the one layer of oracle-bone inscriptions to the "Double-Evidence Method" of Wang Guowei) to undertake research, to develop and to confirm each other. If we can produce results in this respect, then our understanding of the Xia dynasty will become more concrete and more complete. When we reach that point, our discussion of Xia Culture can be said to be "timely." I look forward to that day.[1]

The "Triple-Evidence Method" that Professor Jao mentioned in this talk was not directed solely at Xia Culture, but is a basic principle of all scholarship. Just as Professor Jao indicated parenthetically, his "Triple-Evidence Method" is obviously an extension of the "Double-Evidence Method" (erchong zhengjufa 二重證據法) of Wang Guowei 王國維 (1877-1927). This "Double-Evidence Method" of Wang Guowei can be said to be the twentieth century's most famous research method for research concerning ancient China; in China today there is virtually no scholar who

---

[1] Quoted at Zeng Xiantong 曾憲通, "Xuantang xiansheng san chong zhengjufa qianxi" 選堂先生三重證據法淺析, *Huaxue* 華學 9 (2008): 33.

would not agree with it. It was also in a public lecture that Wang Guowei first mentioned this research method. In 1925, in his class "New Evidence For Ancient History" at Qinghua University's Institute of National Studies, he averred:

> *Today we are lucky to have unearthed sources over and above our paper sources. From these sorts of sources, we can certainly supplement and correct paper sources, and can also confirm that some parts of ancient books are wholly accurate, and that even the coarse sayings of the Hundred Schools express some aspect of the actual situation. This Double-Evidence Method has only become possible now. It can be said with assurance that even those ancient books that have not yet been proven reliable cannot for that reason be negated, while those that have already been confirmed should certainly be treated positively.*[2]

Wang Guowei was a textual critic and paleographer, so that it is not at all surprising that the research method he proposed was limited to written evidence: what he meant by "paper sources" of course pertains to transmitted literature, while his "unearthed sources" referred to unearthed written records such as oracle-bone inscriptions, bronze inscriptions and bamboo and silk manuscripts, and would also include the various manuscripts from Dunhuang even if strictly

---

2 Wang Guowei 王國維, *Gu shi xin zheng* 古史新證, in *Wang Guowei wenji* 王國維文集 (Beijing: Zhongguo wenshi chubanshe, 1997), vol. 4, p. 2.

speaking most of these were "paper sources."[3] Jao Tsung-i's scholarly perspective was even broader, emphasizing material culture—what he termed archaeological remains—as a third type of evidence in addition to transmitted literature and unearthed written records.

Twenty years after Professor Jao first proposed this "Triple-Evidence Method," he further proposed considering two other types of indirect evidence: anthropological sources and ancient historical sources from other countries, making five sorts of evidence in all. This "Quintuple-Evidence Method" shares at least a superficial similarity with what Kwang-chih Chang (1931-2001) had termed the "Five Doors to Shang." In the "Introduction" to his 1980 book *Shang Civilization,* Professor Chang said that there are five types of evidence with which to study Shang civilization, what he termed these five "doors": traditional texts, bronze vessels (inclusive of inscription, vessel type, ornamentation, and handicraft), oracle bones, archaeology, and theoretical models.[4]

In fact, from the beginning of the twentieth century

---

[3] Because of this, Chen Yinke 陳寅恪 qualified the "Double-Evidence Method", saying that it had three types of significance: "the first is to correlate underground artifactual evidence with paper sources"; "second is for ancient texts of our country and those of different nationalities to corroborate each other"; and "third is to compare foreign ideas with the sources we have long had"; see Chen Yinke 陳寅恪, Wang Jing'an xiansheng yishu xu 王靜安先生遺書序 (*Jinmingguan conggao erbian* 金明館叢稿二編) .

[4] Kwang-chih Chang, *Shang Civilization* (New Haven: Yale University Press, 1980) , pp. 1-65. For a Chinese translation, see Zhang Guangzhi 張光直, *Shang wenming* 商文明, Zhang Liangren 張良仁 et al tr. (Shenyang: Liaoning Jiaoyu chubanshe, 2002) .

down to the present, many of China's most authoritative historians have espoused similar "Multiple-Evidence Methods," simply referring to them by different names. As early as 1905, in the Introduction to his *On the Origin of Ancient Government* (*Gu zheng yuanshi lun* 古政原始論), Liu Shipei 劉師培 (1884-1919) proposed a method of correlating books, script and artifacts.[5] Similarly, Fu Sinian 傅斯年 (1896-1950) not only proposed using three types of historical sources to conduct research on ancient history, but he also called these "direct sources." According to Fu's explanation, these direct sources were still insufficient: "If one just sticks to bits of bronze inscriptions and isn't thoroughly familiar with the classics and their commentaries, then you can't hope to be anything other than an engraver of seals." Without historical theory, this would only count as "playing with antiques."[6] Even Gu Jiegang 顧頡剛 (1893-1980), Wang Guowei's adversary in his *New Evidence for Ancient History* speech, also proposed a type of multi-evidence method of historiography, saying: "Today we have received the stimulus of the times and know that we can use the results of archaeology to reconstruct true history, and we can also use anthropological methods to reconstruct mythology and legends; it has finally gradually become possible to bring order to this account which gets messier the

---

5 Liu Shipei 劉師培, "Gu zheng yuanshi lun" 古政原始論, *Guocui xuebao* 國粹學報 1.4（1905）; rpt. In *Liu Shenshu yishu* 劉申叔遺書（Ning Wunan shi jiaoben 寧武南氏校本）, fasc. 18.

6 Fu Sinian 傅斯年, Shixue fangfa daolun 史學方法導論, in *Fu Sinian quanji* 傅斯年全集（Changsha: Hunan Jiaoyu chubanshe, 2003）, p. 310.

more you deal with it."⁷ Most recently, Professor Li Xueqin 李學勤 of Qinghua University has written a blog entitled "The Double-Evidence Method and the Study of Ancient History" in which he has given a summation of this idea: "Decades of historiographical history show that our progress in the field of ancient history relies on the merger of history and archaeology, the corroboration of traditional texts and archaeological discoveries. From now on explorations of the society, economics and thought of the ancient period will continue to proceed in this direction. In fact, in terms of the history of other ancient countries and civilizations, the situation is the same." He concludes by saying "Wang Guowei's 'Double-Evidence Method' was an important contribution to the study of ancient history; now eighty years later it still deserves to be observed and developed".⁸ In sum, regardless of whether we call it a "Double-Evidence Method", "Triple-Evidence Method" or "Quintuple-Evidence Method," not many historians would deny that a multi-evidence method is the mainstream of modern Chinese historiography.

Even though I say that not many historians would deny the importance of a multi-evidence method, nevertheless there are some exceptions; otherwise, today's talk would end right here. The topic of today's talk is some experiences

---

⁷ Gu Jiegang 顧頡剛, "Zixu" 自序, in *Zhongguo shang gu shi yanjiu jiangyi* 中國上古史研究講義 (Beijing: Zhonghua shuju, 1988), pp. 1-2.

⁸ Li Xueqin 李學勤, "'Erchong zhengjufa' yu gu shi yanjiu"「二重證據法」與古史研究, at: http://tongzi.blog.hexun.com.tw/41606555_d.html (posted: 2009-12-6 11:20:01).

證² + 證³ = 證⁵ ≡ 證 ≡ 一

I myself have had with the methodology of historical evidence, which I would like to share with Professor Jao and other colleagues. Fifteen years ago, together with Michael Loewe I co-edited *The Cambridge History of Ancient China*, a large-scale comprehensive history. Although the subtitle of this book is *From the Origins of Civilization to 221 B.C.*, the basic contents only include the historical stages of China's pre-Qin period, which is to say the Shang and Zhou (including the Western Zhou, Spring and Autumn and Warring States periods) for which there are written sources. When Dr. Loewe and I began to plan for this volume, our first basic principle was that every period should have two chapters, one treating the material culture and one treating the literary sources. Thus, for the Shang we had Robert Bagley's "Shang Archaeology" and David N. Keightley's "The Shang: China's First Historical Dynasty"; for the Western Zhou, we had my own "Western Zhou History" and Jessica Rawson's "Western Zhou Archaeology"; for the Spring and Autumn period, we had Lothar von Falkenhausen's "The Waning of the Bronze Age: Material Culture and Social Developments, 770-481 B.C." and Cho-yun Hsu's "The Spring and Autumn Period"; and for the Warring States period we had Mark Edward Lewis's "Warring States Political History" and Wu Hung's "The Art and Architecture of the Warring States Period," all of them arranged together. In the "Introduction" co-authored by Loewe and myself, we especially emphasized that this "multi-evidence method" is not only the mainstream of twentieth century Chinese historiography, but has also been

the traditional method of Western Sinology.[9]

After that book was published, virtually the first book review to be published appeared in China; entitled "After Reading *The Cambridge History of Ancient China*: Who Knows the True Appearance Of Mount Lu?," it was by Professor Xie Weiyang 謝維揚 of the Department of History of Shanghai University.[10] At the beginning of the review, Professor Xie announced that the book had not been the work of Chinese scholars:

> *In all, this book basically has not conscientiously taken into account the great work done in the field of ancient history over the last decade or so, and especially that of Chinese pre-Qin history and archaeology (from this point of view, this book is essentially quite separate from the work of Chinese scholars).*

It is rather hard to understand this criticism. *The Cambridge History of Ancient China* includes a bibliography that is eighty pages long, the great majority of titles cited being works by Chinese scholars. To say that there is some degree of "separation" would of course be acceptable, but to say that the authors of this book have "not conscientiously taken into account" the work of Chinese scholars would seem to be rather excessive. In fact, Professor Xie has still more

---

[9] Michael Loewe and Edward L. Shaughnessy, "Introduction," in *The Cambridge History of Ancient China: From the Origins of Civilization to 221 B.C.* (New York: Cambridge University Press, 1999), see especially pp. 13-14.

[10] Xie Weiyang 謝維揚, "Jianqiao Zhongguo gudai shi du hou: Shei shi Lu shan zhen mianmu" (《劍橋中國上古史》讀後——誰識廬山真面目), *Wenhui bao* 文匯報 7 April 2001.

pointed criticisms, saying that the book's writing style is "contentious":

> Readers will strongly feel that for Chinese scholars, in very many respects this book's writing style is contentious. For instance, this book basically has a negative attitude toward ancient Chinese historical records of texts concerning the period before the Shang dynasty (the title of this book's fourth chapter is "The Shang: China's First Historical Dynasty"). Because of this, although from the time of Sima Qian on, the Xia dynasty has had a fixed place in China's traditional historical system, and although China's scholarly world as a whole considers that the question of Xia history is not at all empty, and has devoted considerable efforts to studying Xia history, <u>The Cambridge History of Ancient China</u> has scrupulously cancelled four hundred full years of Xia-dynasty history (only Kwang-chih Chang's chapter "China on the Eve of the Historical Period" has one brief section of only about one thousand words entitled "The Question of the Xia Dynasty," but Chang's viewpoint obviously does not represent the viewpoint of the book as a whole).

I do not propose today to discuss the important question of the historicity of the Xia dynasty. The reason that *The Cambridge History of Ancient China* began from the Shang dynasty is because we adopted a traditional Western understanding of "history," even if a narrow understanding, that the historical period begins only with the beginning of contemporary written records; the period prior to contemporary written records can only be considered "prehistorical". *The Cambridge History of Ancient China* was published by Cambridge University Press, so it goes without saying that we would use a traditional Western historiographical concept. Xie Weiyang is certainly free to adopt a different concept of history, but to deny completely the work of *The Cambridge History of Ancient China* on the basis of just this would again seem to be excessive; indeed, one might even say "contentious".

There is a theoretical foundation to Xie Weiyang's criticism. Professor Xie said that the "multi-evidence method" (he calls it a "dual method") adopted by *The Cambridge History of Ancient China* belittles traditional Chinese texts.

> *In terms of methodology, given that we have this new understanding of the situation of ancient texts, that The Cambridge History of Ancient China should in this way self-confidently and lightly negate many important ancient historical sources, and before sufficient counter-evidence becomes available, it should not be so cocky as to blithely and completely discard texts and to create its own cookstove. A major feature of the writing of The Cambridge History of Ancient China is its "dual method," which is to treat artifactual evidence and textual evidence in parallel, implying a conception of general distrust toward China's early texts (and because of this it is also dissatisfied with Wang Guowei's "double-evidence method").*

Shortly after this book review by Xie Weiyang was published, he again discussed *The Cambridge History of*

Ancient China's view of historical sources in an article entitled "How Ancient Books Became Books and the Question of the Sources of Ancient History":

> *Perhaps the most recent example of this is <u>The Cambridge History of Ancient China</u>, which appeared not long ago. As a sort of representative work of Western Sinology, as usual it displays an allergic conception toward the question of the sources of ancient history (this can be seen from its lengthy discussion of the question of sources); one noteworthy point is that it has adopted a rarely seen "dual" structure pitting unearthed sources against textual sources. This gives people the impression that: 1) in the face of the complex question of the sources of ancient history, they wish to avoid at all costs a simple expression toward the question of the principles regarding the sources of ancient history; and 2) that at the same time, they intend to use this "dual" method to suggest that their reliance on unearthed artifacts to discuss ancient history is the orthodox viewpoint regarding the sources of ancient history, and that the status of texts correspondingly becomes "dependent." Because of this, even though they have not barefacedly clubbed traditional texts to death, still, based on this style, the place of texts as independent sources for ancient history has become greatly diminished, even to the point of essentially having no independence, so it would be somehow inadequate to rely on texts to discuss ancient history (precisely because of this, even though the textual sources concerning the history of the Xia dynasty are known to everyone, this book refused to include a chapter on the "Xia Dynasty"). It would seem that most contemporary scholars could not approve of this sort of view of*

the sources for ancient history.[11]

I think we can disregard Professor Xie's statement that we wanted "to club traditional texts to death"; how could a group of foreigners have the wherewithal to "club traditional texts to death"? However, when Professor Xie says that one objective of our use of the double-evidence method was to turn traditional texts into dependent evidence and that "most contemporary scholars could not approve" of this sort of double-evidence method, I don't know who most such scholars would be. Would they also not approve of the historiographical conceptions of Liu Shipei, Fu Sinian,

---

[11] Xie Weiyang 謝維揚, "Gu shu cheng shu qingkuang yu gu shi shiliao wenti" 古書成書情況與古史史料問題, in Jilin daxue Guji yanjiusuo ed., *Jin Jingfang jiaoshou bainian danchen jinian wenji* 金景芳教授百年誕辰紀念文集 (Changchun: Jilin daxue chubanshe, 2002). It is evident that the question of greatest concern to Xie is the historicity of the Xia dynasty. Later in this article, he continued:

> This explains why according to the reality of China's early texts, the result of steadfastly holding to a position of positive evidence is unhelpful. In the practice of studying ancient history, the influence of positive evidence needs to be reconsidered. Above I noted that <u>The Cambridge History of Ancient China</u> did not include a chapter on the "Xia Dynasty"; in fact, in essence the geneaology of Xia given in the "Xia Basic Annals" chapter of the Records of the Historian has no basic difference from that of the "Shang Basic Annals" (sic); if we were to backtrack 150 years, then based on the standpoint of the The Cambridge History, it would not be possible to write a chapter on the "Shang Dynasty," because that period wouldn't have the "positive evidence" of the Yinxu oracle bones. However, nowadays nobody doubts the reality of Shang history. Because of this, regarding the question of the "Xia Basic Annals," the standpoint regarding positive evidence to which Cambridge now holds actually reflects a complete refusal to consider the entire and basic expression of China's early texts. In fact, based on the most basic logic, to refute the "Xia Basic Annals" on the basis of positive evidence is very possibly dogmatic. This fact has a very profound significance, which is that we need an appropriate recognition of the basic reasons behind the production of China's early textual editions, a point to which we still need to pay great attention.

Wang Guowei, Li Xueqin, and Jao Tsung-i?

What is strange is that at about the same time there also appeared a criticism of *The Cambridge History of Ancient China* that was diametrically opposed to that of Xie Weiyang. Professor David Schaberg of the University of California at Los Angeles (and now the Dean of Humanities there) also wrote two separate articles discussing *The Cambridge History of Ancient China*, one a very lengthy review article entitled "Texts and Artifacts: A Review of *the Cambridge History of Ancient China*" published in the authoritative sinological journal Monumenta Serica,[12] and the other an address entitled "Some Topics of Debate Concerning Chinese History among Western Scholars over the Last Decade," which he presented at East China Normal University.[13] In the speech that he gave, Schaberg raised the question:

> If in the future there should appear written evidence concerning the Xia dynasty similar to that of the Yinxu oracle-bone inscriptions, there will certainly be scholars who will re-write *The Cambridge History of Ancient China*. Based on the archaeological discoveries of the last fifteen years, I certainly recognize that *The Cambridge History of Ancient China* should already be re-written. However, according to the principles of traditional Western historiography that we adopted, in the continued absence of any contemporary written records, there would still be no need to re-write at least the one point concerning the historicity of the Xia dynasty.

[12] David Schaberg, "Texts and Artifacts: A Review of the *Cambridge History of Ancient China*," *Monumenta Serica* 49 (2001): 463-515.

[13] Shi Jiabo 史嘉柏 (David Schaberg), "Jin shinian Xifang Hanxue jie guanyu Zhongguo lishi de ruogan zhenglun wenti"（近十年西方漢學界關於中國歷史的若干爭論問題：2005年10月27日在華東師範大學的學術演講）, *Haiwai Zhongguoxue pinglun* 海外中國學評論 2 (2007): 47-55.

Below, let us first look at the conflict Edward Shaughnessy of the University of Chicago and Robert Bagley of Princeton University had over the relative values of historical method and archaeological method in the process of writing <u>The Cambridge History of Ancient China</u>. What was the topic of the debate? The archaeologists among the contributors, such as Bagley and Lothar von Falkenhausen to different degrees think that to understand ancient Chinese history one must leave behind the framework of traditional Chinese texts and to establish a new historical interpretation based directly upon the foundation of archaeologically excavated artifacts, and thereafter to use the results of that research to correct and augment the defects of the old texts. On the other hand, the historians among the contributors comparatively emphasize the historicity of the texts, taking ancient historical texts such as the <u>Classic of Documents</u>, <u>Bamboo Annals</u>, <u>Remnant Zhou Books</u> or the <u>Zuo Tradition</u> as their starting point, and then comparing them with other transmitted texts and unearthed texts, even if they cannot render them entirely consistent in the end they can still find consistency among all of the sources. From what I have heard, in the editorial process the two sides criticized one another rather severely, but in the end the two positions were all but unshaken.[14]

As one of the two co-editors of *The Cambridge History of Ancient China*, I am probably clearer than anyone else about the "editorial process" the text went through. It would not be mistaken at all to say that there were "two positions,"

---

[14] Shi Jiabo, "Jin shinian Xifang Hanxue jie guanyu Zhongguo lishi de ruogan zhenglun wenti," p. 55.

but to say that "the two sides criticized one another rather severely" would be more than a little exaggerated. No matter what, all of the authors of *The Cambridge History of Ancient China* agreed with the book's two-fold organization. In the conclusion to his speech, Professor Schaberg drew out a seemingly even-handed choice between these "two positions."

> *At this pivotal moment when sinology is just in the course of a large-scale internationalization, what causes Western scholars to be most suspicious is a very basic question: if we want to understand China's past, to what extent should the primary conceptual models be drawn from models indigenous to Chinese culture? That is to say, should our understanding of ancient Chinese history be completely new, completely scientific, and completely objective? Or should it to the extent possible integrate views that Chinese people have had throughout the ages toward their own past? One definitely must make a choice, it is just that we still do not know what to adopt and what to discard. I believe that engaging in this sort of debate is one of the responsibilities of historians; along with the globalization of Chinese culture, sinologists of all areas should also join together in this debate.*[15]

Who could deny the merits of "completely scientific and completely objective" scholarship? (I am not so sure that "completely new" scholarship is necessarily positive; there is plenty of "old" scholarship that is still worth

---

[15] Schaberg, "Texts and Artifacts," pp. 474-75.

consulting.) However, if we think about this a little more deeply, this choice is not at all even-handed. The problem is that Schaberg has placed "completely scientific, and completely objective" scholarship on one side and "to the extent possible integrate views that Chinese people have had throughout the ages toward their own past" on the other side as mutually exclusive alternatives. Could it be that the views of Chinese people themselves toward the past could not be "completely scientific and completely objective," not to mention "completely new"? In fact, this sort of choice is false. What are the "views that Chinese people have had throughout the ages toward their own past"? According to Schaberg's review of *The Cambridge History of Ancient China*, these are precisely the uses of transmitted texts as historical sources. He says that this type of scholarship is based on "credulity," which is to say that it is not scientific and not objective.

> *For the most part, the texts that historians work with are of late or poorly understood provenance. Either they are known to have been written centuries after the events they recount (as in the case of the Shiji* 史記*) or, worse, they come down to us with no certain information about when and where they were written, or by whom. Even archaeologically recovered writings, like oracle bone and bronze inscriptions, carry with them precious little information about the circumstances of their composition and use, or about the relation of the instance of writing to surrounding historical realities. Further, the rare cases in which archaeology has yielded documents that unambiguously corroborate a received text's account (as in the*

famous case of the Shiji's list of Shang kings) cannot, on any logical grounds, be taken to demonstrate the general accuracy of the received text's narratives. If "traditional history" is defined as historical research based in part on received texts, then it seems necessarily to begin either with radical uncertainty or with credulity. Historians derive both the framework of history (including the chronology of all but a few rulers) and information about its particulars from received texts that are either late or unprovenanced, and they will, therefore, never be able to speak about the narrated past with the precision that characterizes many of archaeology's reconstructions.

The best historians, when they use received texts, routinely treat them as artifacts of dubious provenance. They note the doubts that surround them, cite and critique the best theories about their origins, and then take care not to reach conclusions more precise than the data will support. Bagley's treatment of his evidence, his rejection of unsubstantiated and anachronistic visions of cultural unity, and his minimalist sketch of second millennium cultural connections might serve as a methodological model for historians. ... A received text has, by definition, lost many of the material characteristics that it had when it was first made. It has undergone recopying, editing, compilation— we do not know when, or by whom—and it is practically impossible to reconstruct its first form, whether we imagine it starting as a manuscript on bamboo strips, as a spoken pronouncement, or as a performance in some other medium.[16]

---

[16] Schaberg, "Texts and Artifacts," p. 474-75.

In the conclusion to this book review, Schaberg states clearly his own view: "such texts are presumed less than perfectly reliable, and unsuitable for direct citation as historical authorities."[17]

The positions of Xie Weiyang and David Schaberg could be said to be diametrically opposed, yet both of them were leveling extreme criticisms against one and the same book. Could it be that the book was really so bad? Perhaps it should be asked whether it was really so good. Regardless, I am pleased to take advantage of the opportunity of this "Jao Tsung-i Lecture" to talk about the multiplicity and singularity of historical evidence. As Schaberg has said, transmitted texts are intrinsically of a rather unreliable nature. To be sure, there are some transmitted texts of which "we do not know when, or by whom" they were written. Strictly speaking, all transmitted texts from the pre-Qin era underwent a very long process of transmission before they took their final shape in the Han dynasty. Today in China there is a rather popular "believing antiquity" scholarly movement. This is based in large part on the Warring States, Qin and Han bamboo and silk manuscripts that have been unearthed over the last thirty or forty years; no one could deny that these unearthed texts truly are extremely important. In his famous lecture "Walking Out of the Doubting Antiquity Age", Li Xueqin advocated the view that "the history of scholarship definitely needs to be rewritten":

---

17 Schaberg, "Texts and Artifacts," p. 507

> *No matter how you put it, my idea is that very many of the things that have now been unearthed can be related with transmitted literature; for example, although the <u>Heguanzi</u> 鶡冠子 has not been unearthed, it is so similar to the <u>Huang Di shu</u> 黃帝書 that we can say with certainty not only that it was written by someone from Chu 楚, but also that it was rather early. Similar examples can give us some fixed points from which we can make a number of inferences. The direction is very clear, and that is exactly opposite from the Doubting Antiquity movement.*[18]

Although the direction of bamboo and silk texts that have been unearthed over the last thirty or forty years certainly is "exactly opposite from the Doubting Antiquity movement," that does not mean at all that transmitted texts and pre-Qin texts are entirely the same. Whether it be a matter of some of the texts doubted by the Doubting Antiquity movement, such as the *Heguanzi*, or of the most reliable texts such as the *Classic of Poetry* (*Shi jing* 詩經), the texts that we can now see all underwent editing work by Han-dynasty scholars, and their present state is definitely different from what it originally was. When teaching, I often cite the Preface to the *Examples of Doubtful Meanings in Ancient Books* (*Gu shu yi yi ju li* 古書疑義舉例) of Yu Yue 俞樾 (1821-1907): "To take the printed books of today and to think that they are the true texts of the men of antiquity would be like hearing that bamboo shoots are good to eat and going home and boiling your bamboo sleeping mat." This is of course a joke, but it's

---

[18] Li Xueqin 李學勤, *Zouchu Yigu shidai* 走出疑古時代 (Shenyang: Liaoning daxue chubanshe, 1994), p. 16.

not only a joke; its significance is quite profound. In fact, Li Xueqin himself expressed the same point in his "Walking Out of the Doubting Antiquity Age" lecture.

> *We need to recognize that the appearance of ancient books was very different from what we imagine. I often say that we ought to use our own experience to appreciate the great work of Kong Anguo, or Shu Xi and Xun Xu. When Kong Anguo transcribed texts into modern characters, he didn't really know Warring States script very well and that is why he made lots of mistakes.*[19]

I would venture to say that not only Kong Anguo 孔安國 (c. 156-74 B.C.), Shu Xi 束皙 (261-300) and Xun Xu 荀勖 (d. 289) made "lots of mistakes", but that Liu Xiang 劉向 and Liu Xin 劉歆 also made lots of mistakes. Some of these mistakes have already been pointed out by textual critics over the ages, but the mistakes that no one has yet noticed may be even more serious. To a considerable extent, the reason that pre-Qin texts can be read today is due to the editing work of Han-dynasty scholars, but to a certain extent this editing work necessarily reflects the intellectual background of the Han dynasty. Because of this, it is fine to say that the direction of unearthed bamboo and silk texts is diametrically opposite from the Doubting Antiquity movement, but that doesn't mean that Xie Weiyang's view of their authenticity is certainly to be accepted.

---

[19] Li Xueqin, *Zouchu Yigu shidai*, p. 6.

證² ＋ 證³ ＝ 證⁵ ≡ 證 ＝ 一

If transmitted texts are unreliable, does this mean that we should accept David Schaberg's statement that "to understand ancient Chinese history one must leave behind the framework of traditional Chinese texts and to establish a new historical interpretation based directly upon the foundation of archaeologically excavated artifacts, and thereafter to use the results of that research to correct and augment the defects of the old texts"? I certainly wouldn't want to deny the role that archaeological sources have played in our modern understanding of ancient Chinese history. Just as Li Xueqin said at the very opening of his "Walking out of the Doubting Antiquity Age" lecture: "The use of archaeological discoveries in the study of history is very large. Everyone would admit this point. This point is probably common knowledge." [20] Nevertheless, archaeological evidence itself is not without problems. In the "Introduction" to *The Cambridge History of Ancient China*, in discussing the sources of Chinese history, Michael Loewe and I recognized the value of archaeological sources, but we also pointed out their limitations:

> *Yet, just as in the case of the textual record, the archaeological record is also bound by biases of its own. Only rarely does the evidence result from a deliberate and sustained search, identification, and excavation of a site whose existence is to be inferred from other sources. Thus, the great bulk of what has now been found results from the dual accidents of preservation and discovery. In almost all cases, artifacts had*

---

[20] Li Xueqin, *Zouchu Yigu shidai*, p. 1.

to be buried if they were to survive: above ground they would have been susceptible to the hazards of nature, accidental destruction, and the ravages wrought by man; the choice of articles for burial was subject to differing motives; and of the articles selected for burial, only those of inert substance have usually survived decay. In addition, most of the archaeological work in China is in the nature of salvage, started either thanks to an accidental find or in advance of a basic construction project. The articles that have been found surely represent only a small percentage of those that had been in use in ancient China, and they just as surely derive very largely from the lives of the most privileged members of society; but as they are all that we possess of such evidence, we may well run the risk of exaggerating their importance. In addition, as several of the contributors to this volume point out, archaeology in China is subject to contemporary political and cultural concerns, perhaps chief among which has been the desire to demonstrate the grandeur of Chinese civilization.[21]

Archaeologists are very well aware of these problems, but there are some textual scholars, such as David Schaberg, who seem to have a rather romantic notion of archaeology, often idealizing archaeological sources. If we were to write ancient Chinese history solely on the basis of archaeological sources, I'm afraid that that history would be very impoverished. Schaberg points to Lothar von Falkenhausen as one of his scholarly models. Of course, Falkenhausen is a very fine scholar, but even he could not serve as a model

---

[21] Loewe and Shaughnessy, "Introduction," p. 11.

for everyone. A few years ago, I wrote a review of his book *Chinese Society in the Age of Confucius (1000-250 BC): The Archaeological Evidence,* in the conclusion to which I said:

> *In the introduction and again in the conclusion, [Falkenhausen] explicitly eschews reference to the traditional historical record, insisting that the material record unearthed by archaeologists can give a more objective picture of the period. Such confidence in one's own disciplinary specialization is admirable, but it is also self-defeating. Disciplinary specialization is not, or at least should not be, a competition but rather a cooperative venture in the quest to understand the almost infinite complexities of human society. In Falkenhausen's single-minded concern with the mortuary remains of the period, we learn a great deal about the deaths of the people of the time, but very little about their lives and virtually nothing about their loves. The poetry of the age is missing, and we know that it was an age when poetry (and especially the Poetry) was of surpassing importance. I am glad to learn about the tombs, but I resist the suggestion that they are the keys to open all doors. Indeed, I find that the scholar of ancient society who disregards sources just because they are of a different nature from what he prefers impoverishes his understanding of all sources. ... In pitting the material remains of the age against its textual remains, and in claiming that the material remains are more faithful witnesses to the society of the age, Falkenhausen invites this sort of criticism. In my view, the society or societies of the age or ages of Confucius have only begun to be described in all of their grandeur and complexity, and I very much doubt that any singular disciplinary approach*

is likely to do it or them full justice.[22]

If transmitted texts are problematic and archaeo-logical sources are also problematic in these ways, could it be that the study of history is hopeless, that there is no evidence on which it can rely? Not at all. All evidence, whether transmitted texts, archaeological sources, unearthed written sources, or the theories of various scholarly disciplines or comparisons with other civilizations, has its value and its use, even though it is also not without problems. As long as we recognize its limitations—its intrinsic biases—and as long as we recognize our own inadequacies—which is to say, our own prejudices, we should be able to make a contribution to the study of ancient Chinese history. These contributions will doubtless never reach the point of being completely new, completely scientific, and completely objective. I'm afraid that even Professor Jao Tsung-i could not reach this level. However, as long as we expand our perspectives in the ways that Professor Jao has, not limiting ourselves to any one type of evidence, but rather making use of a double, triple or even more multiple evidential method, we will be able to produce research that is "timely," as Professor Jao has termed it. Professor Jao said that he "looks forward to that day." With him as our model, that day is ever closer.

---

[22] Edward L. Shaughnessy, "Review of Lothar von Falkenhausen, *Chinese Society in the Age of Confucius (1000-250 BC): The Archaeological Evidence*," *Journal of Asian Studies* 66.4 (2007) : 1132.

$證^2 + 證^3 = 證^5 ≡ 證 = 一$

Q & A

After presenting my lecture, the faculty and students present raised several very good questions, many of them revolving about the question of the evidentiary method. Although, like Professor Jao Tsung-i, I too advocated using a multi-evidential methodology to study the cultural history of ancient China, the question is by no means a simple one. As some of the students pointed out, if two types of evidence are inconsistent or even contradictory, how do we decide which evidence to use and which to reject? This is an excellent question, one that I talked around for fifteen or so minutes, discussing historical method in terms of various cases, but in the end I felt like I never did give the students a satisfactory response. I myself was quite dissatisfied, and even though the students were very polite, I suspect that they too must not have been very satisfied. Thus, I hope to take advantage of the publication of this small volume to discuss again this pivotal question, and to see if I can't give a little better response.

In my lecture, I mentioned the "Quintuple Evidence Method" advocated by Professor Jao in his 2003 publication "On the Reconstruction of Ancient History". This "Quintuple Evidence Method" joined "anthropological data" and "ancient history of other countries" to his earlier "Triple Evidence Method"—transmitted literature, unearthed documents, and material artifacts. Professor Jao himself divided these five types of evidence into "direct sources" (the latter three) and "indirect sources" (the first two), suggesting that that the two types of evidence do not have equal evidentiary value. The indirect sources can serve as comparisons, stimulating new ideas, but in the absence of direct sources they cannot serve as independent evidence. In fact, Professor Jao himself intimated that even the three direct sources do not necessarily have equal value:

> *The three types of sources that I have advocated add one type of evidence on to that of Wang Guowei. This is because cultural relics considered as artifacts should be treated separately from the written records of cultural relics. The written records of unearthed artifacts are direct sources even greater value, the importance of which should be especially emphasized.*

Although he never said so in so many words, it seems that his meaning here is that unearthed sources should be regarded as more important than traditional literature, and then, as he did say, unearthed sources "should be treated separately," with written records having "even greater value" than artifacts. If this understanding is not mistaken, then we can rank Professor Jao's various types of evidence according to the following sliding scale:

> *Unearthed written records*
> *Unearthed artifacts*
> *Traditional literature*
> *Anthropological data and comparative history*

I agree that this sort of ranking is generally reasonable, but feel that there is still room for discussion. Let me begin by introducing the views of some of Western Sinology's most authoritative scholars.

Before archaeology had taken over the preeminent place in the study of China's ancient history, Bernhard Karlgren (1889-1978) had already evaluated the relative value of traditional documents. In his article "Legends and cults in ancient China" published in 1946, Karlgren provided a survey of other sinologists' work on ancient China, raising several points of criticism, especially concerning their use of textual evidence. According to Karlgren, ancient Chinese texts should be divided into two general types: "free texts," such as the *Classic of Poetry, Classic of Documents," Zuo Tradition, Discourses of the States, Analects* of Confucius, and *Mencius*, all of which are pre-Qin texts that more or less reflect the historical conditions of the time; and "systematizing texts," such as the *Rites of Zhou, Record of Rites, Mr. Lü's Springs and Autumns, Records of the Historian*, and so on, which were mainly edited in the Han dynasty and which primarily express an idealized view of antiquity, even if they do contain some reliable ancient material. Karlgren argued that such works of his contemporaries as Friedrich Hirth's (1845-1927) *The Ancient History of China* (1908) and Marcel Granet's (1884-1940) *Danses et légends de la Chine ancienne* (1926) were largely based on "systematizing texts" and were therefore largely unreliable. Even *La Chine antique* (1927) of Henri Maspero (1882-1945) did not entirely escape this methodological flaw. Even though Karlgren's evidential method seems not to have gained any notice in China, among Western sinologists it has had a very great influence.[23]

By the 1970s, some Western sinologists such as David N. Keightley had gone a step further even than Karlgren and argued that literary records should be divided into two types: "hard" and "soft." Keightley is an oracle-bone specialist, and has spent his career researching the history of the Shang dynasty. According to him, all traditional written sources, regardless of whether they were those termed by Karlgren as "free texts" or "systematizing texts" are soft evidence and should not be used as independent evidence for the cultural history of ancient China; only unearthed written records should be regarded as hard evidence. Because of this, he argued that Shang history should be written only on the basis of Shang oracle-bone inscriptions. This sort of "hard" methodology has won the praise of many Western scholars. For example, in my lecture I mentioned that David Schaberg had written a review of *The Cambridge History of Ancient China*. In his review, Schaberg singled out Keightley as a model for Western historians for the caution with which he uses evidence, and suggested that other historians should also use only this sort of "hard" evidence.[24]

---

23 Bernhard Karlgren, "Legends and cults in ancient China," *Bulletin of the Museum of Far Eastern Antiquity* 18 (1946) : 199-365。Goren Malmqvist makes the point that this is the only major study written by Karlgren that has never been translated into Chinese; see N.G.D. Malmqvist, *Bernhard Karlgren: Portrait of a Scholar* (Bethlehem, Penn.: Lehigh University Press, 2011), p. 229.

24 Schaberg, "Texts and Artifacts", pp.476-77.

證² + 證³ = 證⁵ ≡ 證 = 一

Among Western scholars there are also those, such as *The Cambridge History of Ancient China* contributors Robert W. Bagley and Lothar von Falkenhausen, who hold that written records, including both traditional literature and also unearthed records, are not as objectively reliable as archaeologically excavated material culture. In my lecture, I already discussed Falkenhausen's views and so will not repeat myself here.

My own view is more or less indecisive. I don't have a single rule or method. If I absolutely had to propose a relative ranking for evidence, it would probably closely resemble that of Jao Tsung-i.

> *Unearthed written records*
> *Unearthed artifacts*
> *Traditional literature*
> *Anthropological data and comparative history*

Likewise, I admire David Keightley's historiographical caution. However, while I feel that such caution is perhaps appropriate for Shang history, I am not so sure that it is a model for all periods or places. There are few transmitted sources for the Shang dynasty (in fact, there may be none, the various documents in the "Shang Documents" section of the *Classic of Documents* possibly all having been written in the Zhou dynasty, even if they might contain some Shang-dynasty materials), so that historians of the period can rely exclusively on oracle-bone inscriptions without sacrificing much evidence. However, the situation is completely different with regard to the cultural history of the Zhou dynasty, not to mention that of the Qin and Han dynasties, and such a historiographical method would be totally inappropriate for them. Of course, one could study the history of the Western Zhou based exclusively on oracle-bone and bronze inscriptions. However, as I pointed out in my earlier criticisms of Falkenhausen, to do so would be to miss the heart of that period. To study the cultural history of the Zhou dynasty exclusively on the basis of unearthed sources would be overly cautious. To do this sort of study, we really do need a multi-evidence methodology, which is to say that we need to use unearthed written records and unearthed artifacts to be sure, but to use also transmitted literature.

Having said this, I fear that I still have not given my questioners a satisfactory response. They asked how to choose between two types of evidence that are inconsistent or even contradictory. I feel that in making this sort of choice, we should consider not only why one type of evidence is reliable, but also why the other type is not reliable, and try to demonstrate when and why the unreliable evidence may have been created. Let me use something that I know a little bit about, the chronology of the Western Zhou dynasty, as a simple example. Most traditional histories, such as the *Tongjian wai ji* 通鑑外紀, *Huangji shi jing* 皇極經世, the *Tong kao* 通考 etc., give the length of reign of King Zhao of Zhou 周昭王 as 51 years. However, the *Bamboo Annals*, which was discovered in A.D. 279, records that King Zhao died after only 19 years in power. Most contemporary historians accept the Bamboo Annals record as reliable, and there are even some Western Zhou bronze inscriptions that seem to corroborate it. However, it is worth considering also where the number of 51 years came from. I have previously suggested that this number is very possibly derived from several other bits of data concerning Western Zhou chronology: that the reign of King Mu of Zhou 周穆王 began one hundred years after the beginning of the dynasty; that King Wu of Zhou 周武王 died two years after his conquest of the Shang dynasty; that the Duke of Zhou 周公 served as regent for seven years; and that Kings Cheng 成王 and Kang 康王 enjoyed forty years when "punishments were not used." If we add the latter three of these figures together (2+7+40=49), and subtract that from the 100 years to the beginning of the reign of King Mu (100-49=51), we arrive at a 51-year reign for King Zhao. This suggests that the transmitted figure was derived and therefore has no value as primary evidence is not at all strange, and is consistent with standard historical method. This does not necessarily mean that the unearthed evidence—the Bamboo Annals—is reliable, but for the present it would seem to be preferable.

If I can give another example, we will see that the principle is not necessarily always so simple. I just mentioned that the traditional view is that King Wu of Zhou died two years after the conquest of Shang. This view was originally recorded in the "Jin teng" 金縢 chapter of the *Classic of Documents*, and was later incorporated into both the "Fengshan shu" 封禪書 and "Zhou benji" 周本紀 chapters of the *Records of the Historian*. Although the

"Ming tang" 明堂 chapter of the *Remnant Zhou Documents* (*Yi Zhou shu* 逸周書) has a different figure, that "King Wu died six years after having conquered the Shang dynasty," manifestly different from the record of the "Jin teng" chapter, most historians regard the *Classic of Documents* as more reliable than the *Remnant Zhou Documents*, and thus have accepted that King Wu died two years after the conquest. Despite this, a few years ago Tsinghua University published a Warring States bamboo-strip manuscript entitled *The Record of King Wu Being Ill and the Duke of Zhou Offering Himself as a Substitute for the King*, the contents and language of which are essentially identical with those of the "Jin teng" chapter of the *Classic of Documents*; this manuscript should be regarded as the earliest version of that text. One difference between the Tsinghua manuscript and the received text of the "Jin teng" chapter is that the opening of the manuscript clearly reads "three years after King Wu conquered the Shang," clearly different from the received text. Based on the principle that accords precedence to unearthed written records, we should surely accept the evidence of the Tsinghua manuscript and suppose that King Wu did not die until at least three years after the conquest of Shang. Would this not entail rejecting not only the "Ming tang" chapter of the *Remnant Zhou Documents*, but also the received "Jin teng" chapter? Not necessarily. Even though in other ancient texts there is some support for King Wu dying in the third year after the conquest (such as the "Yao lüe" 要略 chapter of the *Huainanzi* 淮南子), I still do not think this is decisive. In classical Chinese, the use of numbers was more or less confused, so that under certain conditions "two years" (*er nian* 二年) could be the same as "three years" (*san nian* 三年), which is to say that if something occurred "two years after" such-and-such an event, it could also be said to have occurred in "the third year" after it. However, this is not true in the opposite direction, by which I mean that if something happened "three years after" something else, there is no way that it could be construed as happening "two years after". Of course, we could simply assume that the received text of the "Jin teng" chapter reflects a copyist's error that entered into the text at some point during its long process of copying and re-copying, but I wonder if such an explanation is not a bit too conjectural.

I don't know whether these examples can explain any principle. Perhaps they explain simply that the questions raised by the University of Hong Kong students are extremely good questions—which is to say extremely difficult questions. We can give simple answers to them, but once we try to use them in doing historical research we almost invariably meet with various kinds of exceptions. By way of summary, perhaps I can do no better than to return to the title of the lecture: there are multiple types of evidence, and these multiple types of evidence can all be used, but only if all types of evidence have their own intrinsic value. Some evidence can serve as direct proof, some evidence as indirect proof, while other evidence can serve as counter-proof. As long as we are clear about the background and the purpose for which it was created, all evidence has its uses.

夏含夷教授簡介

夏含夷（Edward L. Shaughnessy）是美國芝加哥大學東亞語文與文明系顧立雅講座教授。主要研究領域為西周及戰國時期的文化史、古文字學、《易經》等。重要著作包括《西周史料：銅器銘文》（1991）、《易經：馬王堆帛書本的首次英譯》（1996）、《溫故知新錄：商周文化史管見》（1997）、《孔子之前：中國經典的創造研究》（1997，2013中文譯本）、《古史異觀》（2005）、《重寫中國古代文獻》（2006，2012中文譯本）、《興與象：中國古代文化史論集》（2012）、《出土之易》（2014）等，與魯惟一（Michael Loewe）合編了《劍橋中國古代史：從文明的起源到公元前221年》（1999），也曾任著名漢學雜誌《古代中國》（Early China）主編。

About Prof. Edward L. Shaughnessy:

Edward L. Shaughnessy is Creel Distinguished Service Professor of Early China in the Department of East Asian Languages and Civilizations of the University of Chicago. His major fields of research are Western Zhou and Warring States cultural history, paleography, and the *Classic of Changes*. His major publications include *Sources of Western Zhou History: Inscribed Bronze Vessels* (1991), *I Ching: The Classic of Changes, The First English Translation of the Newly Discovered Second-Century B.C. Mawangdui Texts (1996)*, *A Record of Reanimating the Old and Knowing the New: Studies in Shang and Zhou Cultural History* (1997, in Chinese), *Before Confucius: Studies in the Creation of the Chinese Classics* (1997, Chinese translation 2013), *A Different View of Ancient History* (2005, in Chinese), *Rewriting Early Chinese Texts* (2006, Chinese translation 2012), *Arousal and Image: A Collection of Essays on Ancient Chinese Cultural History* (2012, in Chinese), and *Unearthing the Changes* (2014). He was the co-editor (with Michael Loewe) of *The Cambridge History of Ancient China: From the Origins of Civilizations to 221 B.C.* (1999), and also served as the editor of the journal *Early China*.

TEACHING AND SCHOLARLY SERVICE

1984-present: Department of East Asian Languages and Civilizations, U. of Chicago; 2006-present: Lorraine J. and Herrlee G. Creel Distinguished Service Professor of Early China 1997-2006: Lorraine J. and Herrlee G. Creel Professor of Early China; 1996-1997: Professor; 1990-1996: Associate Professor; 1984-1990: Assistant Professor; 1990-1993, 2008-2012: Chair.

2006-present: Creel Center for Chinese Paleography, Director.

1988-1996: *Early China*, Editor.

Editorial Board Member: *Guowen xuebao* 國文學報 (Taiwan), *Guji zhengli yanjiu xuekan* 古籍整理研究學刊 (China), *Jian bo* 簡帛 (China), *Hanzi yanjiu* 漢字研究 (Korea)

Advisory Board Member: University of Notre Dame, Asian Studies Advisory Board, Chinese University of Hong Kong, Institute of Chinese Studies Advisory Board

EDUCATION

1978-1983: Stanford University (Stanford, CA) M.A. 6/80, Ph.D. 6/83 (Diss. "The Composition of the *Zhouyi*"), Asian Languages.

1977-1978: Kyoto Japanese Language Center (Kyoto, Japan).

1974-1977: Tiande Academy 天德黌舍 (Taipei, Taiwan); independent study of Chinese classics with Aisingioro (Liu) Yu-yun 愛新覺羅毓鋆.

9/70-5/74: University of Notre Dame (South Bend, IN) B.A. 5/74, Theology.

PROFESSIONAL AWARDS AND APPOINTMENTS

2013.6: College de France: Visiting Professor
2013.1-4: Institute of Chinese Studies, Chinese University of Hong Kong: Distinguished Visiting Professor
2008.4: École Pratique des Hautes Études: Visiting Professor
2007-08: National Endowment for the Humanities: Research Award
2002-03: J. William Fulbright Foreign Scholarship: Program for Research in China
1994: Center for Chinese Studies, Taipei, Taiwan
1993: CSCC, National Academy of Sciences: Program for Research in China
1984-85: Andrew W. Mellon Fellowship for Chinese Studies
1983-84: CSCPRC, National Academy of Sciences, Program for Research in China
1982-83: Mabelle McLeod Lewis Memorial Award
1981-82: Mrs. Giles Whiting Foundation Award

PUBLICATIONS

Monographs

Unearthing the Changes: *Recently Discovered Manuscripts of and Relating to the Yi Jing*. New York: Columbia University Press, 2014.
Xing yu xiang: Zhongguo gudai wenhua shi lunwenji 興與象：中國古代文化史論文集 (*Arousals and Images: Essays on Ancient Chinese Cultural History*).

Shanghai: Shanghai Guji chubanshe, 2012.

Chinese Wisdom: Philosophical Insights from Confucius, Mencius, Laozi, Zhuangzi and Other Masters. London: Duncan Baird Publishers, 2010; American edition entitled *Confucian & Taoist Wisdom: Philosophical Insights from Confucius, Mencius, Laozi, Zhuangzi and Other Masters.*

Yuanfang zhi shi xi: Gudai Zhongguo *jingxuanji* 遠方之時習：古代中國精選集 (*Timely practices from distant parts: Selections from Early China*), editor. Shanghai: Shanghai Guji chubanshe, 2008.

Rewriting Early Chinese Texts. Albany, NY: SUNY Press, 2006. Chinese translation: *Chongxie Zhongguo gudai wenxian* 重寫中國古代文獻. Tr. Zhou Boqun 周博群. Shanghai: Shanghai Guji chubanshe, 2012.

Ancient China: Life, Myth and Art. London: Duncan Baird Publishers, 2005. Polish translation: *Chiny: Życie, Legendy I Sztuka.* Warzawa: National Geographic Society, 2005; French translation: La Chine ancienne: Vie, art et mythes, tr. Emmanuel Pailler. Paris: Gründ, 2005; Spanish translation: *La Antigua China: Vida, Mitologia y Arte.* Madrid: Ediciones Jaguar, 2005; Portugese translation: Antiga China. Lisbon: *Edição Única*, 2005.

Gu shi yi guan 古史異觀 (*A different view of ancient history*). Shanghai: Shanghai Guji chubanshe, 2005.

China: Land of the Heavenly Dragon, general editor. London: Duncan Baird Publishers, 2000. American edition: *China: Empire and Civilization.* New York: Oxford University Press, 2000. Translated into

French, German, Polish, Russian, Czech, Estonian, Danish and Slovak.

The Cambridge History of Ancient China: From the Origins of Civilization to 221 B.C., edited with Michael Loewe. New York: Cambridge University Press, 1999. Translated into Korean.

Before Confucius: Studies in the Creation of the Chinese Classics. Albany, N.Y.: SUNY Press, 1997. Chinese translation: Kongzi zhi qian 孔子之前. Huang Shengsong 黃聖松, Yang Jixiang 楊濟襄, Zhou Boqun 周博群 *et al trans.*, Fan Limei 范麗梅 and Huang Kuanyun 黃冠雲 ed. Taipei: Wanjuanlou, 2013.

Wen gu zhi xin lu: Shang Zhou wenhua shi guanjian 溫故知新錄：商周文化史管見 (*A record of reanimating the ancient and knowing the new: Views of Shang and Zhou cultural history*). Taipei: Daohe chubanshe, 1997.

New Sources of Early Chinese History: An Introduction to the Reading of Inscriptions and Manuscripts, editor. Berkeley: Institute of East Asian Studies, University of California, Berkeley, and the Society for the Study of Early China, 1997. Chinese Translation: *Zhongguo guwenzixue daolun* 中國古文字學導論 (*Introduction to Chinese paleography*). Shanghai: Zhong Xi shuju, 2013.

I Ching, The Classic of Changes: The First English Translation of the Newly Discovered Second-Century B.C. Mawangdui Texts. New York: Ballantine Press, 1997.

Sources of Western Zhou History: Inscribed Bronze Vessels. Berkeley: University of California Press, 1991.

Ritual and Reverence: Chinese Art at the University of Chicago, with Robert Poor, Harrie A. Vanderstappen, Richard A. Born and Sue Taylor. Chicago: Alfred A. Smart Gallery of Art, 1989.

A Concordance of the Xiaotun Nandi Oracle-Bone Inscriptions, with Cai Fangpei and James F. Shaughnessy, Jr. Early China Special Monograph Series, Number One. Chicago, July, 1988.

Essays

"Xifang Hanxuejie li de liangwei Zhongguo jiandu xue dashi" 西方漢學界裏的兩位中國簡牘學大師 (Western Sinology's two great scholars of Chinese bamboo and wooden strips). *In Fudan daxue Chutu wenxian yu guwenzi yanjiu zhongxin shi nian huadan lunwenji* 復旦大學出土文獻與古文字研究中心十年華誕論文集 (*Essays celebrating the tenth anniversary of the establishment of the Center for the Study of Unearthed Texts and Paleography of Fudan University*). (Shanghai, 2015).

"Xifang jinwen dashi: Weng Youli (Ulrich Unger, 1930-2006) xiaozhuan" 西方金文大師：翁有理 (Ulrich Unger, 1930-2006) 小傳 (A great Western scholar of bronze inscriptions: A short biography of Ulrich Unger [1930-2006]). *Guwenzi yanjiu* 古文字研究 (*Paleographic Research*) 30 (2014): 614-19.

"Xifang Hanxuejia Zhongguo guwenzi yanjiu gaiyao" 西方

漢學家中國古文字研究概要 (An outline of Western Sinologists' research on Chinese paleography). *Jianbo* 簡帛 (*Bamboo and silk*) 9 (2014): in press.

"Qi yu jiagu: Xifang Hanxuejia Shang Zhou jiaguwen yanjiu gaiyao" 契于甲骨：西方漢學家商周甲骨文研究概要 (Inscribed on shell and bone: An outline of Western Sinologists' research on Shang and Zhou oracle-bone inscriptions). In *Jiaguwen yu Yin Shang shi* 甲骨文與殷商史 (*Oracle-Bone Inscriptions and Shang History*), ed. Song Zhenhao 宋鎮豪 (Beijing: Shehui kexueyuan chubanshe, 2014), in press.

"Philosophy or Bamboo: The Reading and Writing of Warring States Manuscripts." *China Reviews International* 19.2 (2014): 199-208.

" 'Xia Shang Zhou duandai gongcheng' shi nian hou zhi pipan: Yi Zhou zhu wang zai wei niandai wei lizheng" 「夏商周斷代工程」十年後之批判：以西周諸王在位年代為例證 (The Xia-Shang-Zhou Chronology Project Ten Years Later: Taking the Reign Dates of the Western Zhou Kings as Evidence). In *Di si jie Guoji Hanxue huiyi lunwenji: Chutu cailiao yu xin shiye* 第四屆國際漢學會議論文集：出土材料與新視野 (*Essays from the fourth international conference on Sinology: Unearthed materials and new perspectives*). Taipei: Academia Sinica, 2013. pp.341-380.

"Jinian xingshi yu shi shu zhi qiyuan" 紀年形式與史書之起源 (The annals form and the beginnings of historical writing). In *Jian bo, jingdian gu shi* 簡帛、經典、古史 (Bamboo and silk, classics and ancient history). Ed. Chen Zhi 陳致. (Shanghai: Shanghai Guji chubanshe, 2013). Pp. 39-46. Revised and expanded English version: "The Qin *Bian Nian Ji* 編年記 and the Beginnings of Historical Writing in China." Minneapolis Institute of Art

"A Special Use of the Character 鄉 in Oracle-Bone Inscriptions and Its Significance for the Meaning of Early Chinese Divination: With Comments on the First Line of the *Yi Jing*." In *Institute of Chinese Studies Visiting Professor Lecture Series (III)*. *Journal of Chinese Studies* Special Issue. Hong Kong, 2013. pp. 163-77.

"The Zhou Dynasty and the Birth of the Son of Heaven." In Maria Khayutina ed. Qin: *The Eternal Emperor and His Terrcotta Warriors*. Bern: Bernisches Historisches Museum, 2013. pp. 17-26.

"La dinastia Zhou." In *La Cina*, Vol 2: *Dall'età del Bronzo all'impero Han*. Ed. Tiziana Lipiello and Maurizio Scarpari. Torino: Einaudi, 2013. pp. 77-133.

"Paleography." Oxford Bibliographies Project. Oxford University Press. 2013.

"Mu tianzi zhuan yu Mu wang shidai tongqi" 穆天子傳與穆王時代銅器 (The *Mu tianzi zhuan* and King Mu Period Bronzes), *Rao Zongyi guoxueyuan yuankan* 饒宗頤國學院院刊 (*Bulletin of the Jao Tsung-I National Studies Institute*). English translation: "The Mu tianzi zhuan and King Mu Period Bronzes," *Rao Zongyi Guoxueyuan yuankan* 饒宗頤國學院院刊 (*Bulletin of the Jao Tsung-i National Studies Institute*) 1 (2014): 55-75.

"Xian Qin shidai 'shu' zhi chuanshou: Yi Qinghua jian *Zhai Gong zhi gu ming* wei li" 先秦時代「書」之傳授：以清華簡〈祭公之顧命〉為例 (The transmission of 'documents' in the Pre-Qin Period: Taking the Qinghua strip *Zhai Gong zhi gu ming* as an example). *Qinghua jian yanjiu* 清華簡研究 (*Researches on the Qinghua strips*) 1 (2012): 217-27.

"Fei chang dao kao" 非常道考 (Inconstant ways), *Guoxue xuekan* 國學學刊 (*National studies journal*) 2011.4: 39-45.

"Shilun Xiang zi zai He zu buci li yizhong teshu yongfa" 試論鄉字在何組卜辭裏一種特殊用法 (On a special usage of the character *Xiang* 鄉 in the He-Group Oracle-Bone Inscriptions), in *Xing yu xiang: Zhongguo gudai wenhua shi lunwenji* 興與象：中國古代文化史論文集 (*Arousals and Images: Essays on Ancient Chinese Cultural History*). Shanghai: Shanghai Guji chubanshe, 2012

"Si lun *Zhushu jinian* de cuojian zhengju" 四論《竹書紀年》的錯簡證據 (A fourth discussion of the evidence for misplaced strips in the Bamboo Annals), in *Xing yu xiang: Zhongguo gudai wenhua shi lunwenji* 興與象：中國古代文化史論文集 (*Arousals and Images: Essays on Ancient Chinese Cultural History*). Shanghai: Shanghai Guji chubanshe, 2012

"He wei 'zhong zhong xueshu'?" 何謂「中中學術」？ (What is 'China-Centered Scholarship), in *Xing yu xiang: Zhongguo gudai wenhua shi lunwenji* 興與象:中國古代文化史論文集 (*Arousals and Images: Essays on Ancient Chinese Cultural History*).

Shanghai: Shanghai Guji chubanshe, 2012

"'Xing' yu 'Xiang': Jian lun zhanbu he shige de guanxi ji qi dui Shi jing he Zhou Yi de xingcheng zhi yingxiang" 「興」與「象」：簡論占卜和詩歌的關係及其對《詩經》和《周易》的形成之影響 ('Arousal' and "Image": A simple discussion of the relationship between divination and song and of its influence on the formation of the Classic of Poetry and the Zhou Changes), Luojia jiangtan 珞珈講壇 (Luojia Forum) 6 (2011): 71-89. Translated into Korean as "Imiji pulro ilukigi: Kodae Chungguk Yok kwa Si ui sang'gwansong," in *Hwa i pudong ui Tongasiahak: Minjoksa wa kodae Chungguk yon'gu charyo songch'al (East Asian Studies Harmonized and yet Different: Reflections on the National History and the Sources of Studying Early China)*, edited by Jae-hoon Shim (Seoul: P'urun yoksa, 2012), pp.135-163.

"Fuyang Han jian Zhou Yi jiance xingzhi ji shuxie geshi zhi lice" 阜陽漢簡《周易》簡冊形制及書寫格式之蠡測 (Estimating the nature and written form of the Fuyang *Zhou Changes* bamboo strips), *Jianbo* 簡帛 (*Bamboo and silk*) 6 (2011): 427-36.

"Cong Zuoce Wu he zai kan Zhou Mu Wang zai wei nianshu ji niandai wenti" 從作冊吳盉再看周穆王在位年數及年代問題 (From the Zuoce Wu he looking once again at the length of reign and dates of King Mu of Zhou), in Zhu Fenghan 朱鳳瀚 ed., *Xinchu jinwen yu Xi Zhou lishi* 新出金文與西周歷史 (*Newly appearing bronze inscriptions and Western Zhou history*) (Shanghai: Shanghai Guji chubanshe, 2011), pp. 71-78.

"History and Inscriptions, China," in *The Oxford History of Historical Writing*, Volume 1: *Beginnings to AD 600*, General Editor Daniel Woolf, Volume Editors Andrew Feldherr and Grant Hardy (Oxford: Oxford University Press, 2011), pp. 371-93.

"Of Riddles and Recoveries: The *Bamboo Annals*, Ancient Chronology, and the Work of David Nivison." *Journal of Chinese Studies* 52 (2011): 269-90.

"*Zhou Yi 'Yuan heng li zhen*' xin jie: Jianlun Zhou dai xi zhen xiguan yu *Zhou Yi* gua yao ci de xingcheng" 《周易》「元亨利貞」新解——兼論周代習貞習慣與《周易》卦爻辭的形成 (A new explanation of the phrase "*Yuan heng li zhen*" in the Zhou Changes: Together with a discussion of the practice of repeat divination in the Zhou period and the formation of the hexagram and line statements of the *Zhou Changes)*, *Zhou Yi yanjiu* 周易研究 (*Studies of the Zhou Changes*) 2010.5: 3-15.

"You Meixian Shan shi jiazu tongqi zai lun Shanfu Ke tongqi de niandai: Fudai zai lun Jin Hou Su bianzhong de niandai" 由眉縣單氏家族銅器再論膳夫克銅器的年代：附帶再論晉侯蘇編鐘的年代 (A new discussion of the dates of the Shanfu Ke bronze vessels from the perspective of the Meixian Shan family bronzes: With an appended renewed discussion of the date of the Jin Hou Su bell set), in *Zhongguo gudai qingtongqi guoji yantaohui lunwenji* 中國古代青銅器國際研討會論文集, ed. Shanghai bowuguan and Xianggang Zhongwen daxue Wenwuguan

(N.p., 2010), pp. 165-78. Translated into Korean as "Mihyon Sonssi kajok ch'ongdonggi rul t'onghan Sonbu Kuk ch'ongdonggi yondae chaegoch'al: Jinhu So p'yonjong yondae wa kwalryon hayo," in *Hwa i pudong ui Tongasiahak: Minjoksa wa kodae Chungguk yon'gu charyo songch'al (East Asian Studies Harmonized and yet Different: Reflections on the National History and the Sources of Studying Early China)*, edited by Jae-hoon Shim (Seoul: P'urun yoksa, 2012), pp.207-232.

"The Beginnings of Writing in China," in *Visible Language: Inventions of Writing in the Ancient Middle East and Beyond* (Chicago: The Oriental Institute, 2010), pp. 215-24.

"La Dinastia Zhou," in Cina, Volume *1.2: Dall'età del Bronzo al Primo Impero,* ed. Prof. Tiziana Lippiello and Prof. Maurizio Scarpari (Torino: Einaudi, 2010), in press.

"Zai lun Zhouyuan bu ci si zi yu Zhou dai bu shi xingzhi zhu wenti" 再論周原卜辭囟字與周代卜筮性質諸問題 (A second discussion of the Zhouyuan oracle-bone character *si* as well as such questions as the nature of Zhou dynasty turtle-shell and milfoil divination), in *2007 Zhongguo jianbo xue Guoji luntan lunwenji* 2007 中國簡帛學國際論壇論文集, ed. Guoli Taiwan daxue Zhongguo wenxue xi (Taipei: Guoli Taiwan daxue Zhongguo wenxue xi, 2011), 17-47.

"Chongxie Rujia jingdian: Tantan zai gudai xieben wenhua zhong chaoxie de quanshi zuoyong" 重寫儒家經典：談談在古代寫本文化中抄寫的詮釋作用 (Rewriting the Confucian classics: On the hermeneutical function of copying in the manuscript culture of antiquity), in *Jingdian de xingcheng, liuchuan yu quanshi* 經典的形成、流傳與詮釋 (*The formation, transmission and hermeneutics of the classics*), vol. 2, ed. Lin Qingzhang 林慶彰 and Jiang Qiuhua 蔣秋華 (Taipei: Taiwan Xuesheng shuju, 2008), in press.

"Zai shuo Xici Qian zhuan zhi, Kun xi pi" 再說〈繫辭〉乾專直坤翕闢 (Once again on the *Appended Statements'* Qian's being curled and straight, Kun's being closed and open), *Wenshi* 文史 91 (2010.2): 273-75.

"Arousing Images: The Poetry of Divination and the *Divination of Poetry*", in *Divination and Interpretation of Signs in the Ancient World*, ed. Amar Annus. Oriental Institute Seminars 6 (Chicago: The Oriental Institute of the University of Chicago, 2010), pp. 61-75.

"Cong chutu wenzi ziliao kan Zhou Yi de bianzuan" 從出土文字資料看《周易》的編纂 (Looking at the composition of the *Zhou Changes* on the basis of excavated written materials), in 2009 *Zhou Yi jing zhuan wenxian xin quan* 周易經傳文獻新詮 (*2009 New hermeneutics on the Zhou Changes classic and commentary texts*), ed. Zheng Jixiong 鄭吉雄 (Taipei: Taiwan daxue chuban zhongxin, 2010), 34-49.

"Lunar-Aspect Terms and the Calendar of China's Western Zhōu Period," in Time and *Ritual in Early China*, Thomas O. Höllmann & Xiaobing Wang-Riese ed. (Wiesbaden: Harrassowitz, 2009), pp. 15-32.

"Jianlun 'Yuedu xiguan': Yi Shangbo *Zhou Yi Jing* gua wei li" 簡論「閱讀習慣」：以上博《周易‧菉》卦為例 (A brief discussion of "Reading Practices: Taking Jing hexagram of the Shanghai Museum Zhou Yi as an example), *Jianbo* 簡帛 (*Bamboo and silk*) 4 (2009): 385-94.

"Chronologies of Ancient China: A Critique of the 'Xia-Shang-Zhou Chronology Project," in *Windows on the Chinese World: Reflections by Five Historians*, ed. Clara Wing-chung Ho. Lanham, Md.: Lexington Books, 2008. Pp. 15-28.

"Gongyuanqian 1000 nian qianhou dong xi wenming jiaoliu san ze" 公元前1000年前後東西文明交流三則 (Three studies of East West cultural exchange at about 1000 B.C.), *Hua xue* 華學 (*Sinology*) 9 (2008): 288-90.

"San lun *Zhushu jinian* de cuojian zhengju" 三論《竹書紀年》的錯簡證據 (A third discussion of evidence for misplaced strips in the *Bamboo Annals*), *Jianbo* 簡帛 (*Bamboo and silk*) 3 (2008): 403-415.

"Changing Semiotics and Hermeneutics: A Review of Zheng Jixiong 鄭吉雄, *Yi tuxiang yu Yi quanshi* 易圖象與易詮釋," *China Reviews International*, 14.1 (Spring 2007): 312-17.

"The Bin Gong Xu Inscription and the Beginnings of the Chinese Literary Tradition," in T*he Harvard-Yenching Library 75th Anniversary Memorial Volume*, ed. Wilt Idema. Hong Kong: The Chinese University Press, 2007. Pp. 1-19.

"The Religions of Ancient China," *in A Handbook of Ancient*

Religions, ed. John Hinnells. Cambridge, England: Cambridge University Press, 2007. Pp. 490-536.

"You tongqi mingwen chongxin yuedu *Shi Da Ya Xia Wu*" 由銅器銘文重新閱讀《詩‧大雅‧下武》 (A new reading of the *Shi jing* poem *Xia Wu* on the basis of bronze inscriptions), in *Qu Wanli xiansheng bai sui danchen guoji xueshu yantaohui lunwenji* 屈萬里先生百歲誕辰國際學術研討會論文集 (*Essays from the International Conference Commemorating the Hundreth Anniversary of Mr. Qu Wanli's Birth*). Taibei: Taiwan daxue, 2007. Pp. 65-69.

"Shilun Shangbo Zhou Yi de gua xu" 試論上博《周易》的卦序 (A trial discussion of the hexagram sequence in the Shanghai Museum *Zhou Changes*), *Jianbo* 簡帛 (*Bamboo and silk*) 1 (2006): 97-105.

"Shi zhong: Jianlun Mao Shi de xungu fangfa yi ze" 釋潀：兼論毛詩的訓詁方法一則 (An explanation of the character 潀: With a discussion of exegetical methods as applied to the *Shi jing*), *Zhonghua wenshi luncong* 中華文史論叢 (*Essays on Chinese cultural history*) 2006.3: 77-85.

"Shilun Xi Zhou tongqi mingwen de xiezuo guocheng: Yi Meixian Shan shi jiazu tongqi wei li" 試論西周銅器銘文的寫作過程：以眉縣單氏家族銅器為例 (A trial discussion of the process of composition of Western Zhou bronze inscriptions: Taking the Meixian Shan Family bronzes as examples), in *Chutu wenxian yu Zhongguo sixiang yantaohui lunwenji* 新出土文獻與先秦思想重構論文集 (*Collection of papers reconsidering newly excavated documents*

and Pre-Qin thought), ed. Guo Lihua 郭梨花. Taipei: Taiwan Guji chuban youxian gongsi, 2007. Pp. 119-130; English version published as "The Writing of a Late Western Zhou Bronze Inscription," *Asiatische Studien/Études Asiatique*s LXI.3 (2007): 845-877.

"Cong Lu Gui kan Zhou Mu Wang zai wei nianshu ji niandai wenti" 從親簋看周穆王在位年數及年代問題 (A look at the problems of the length and date of the reign of King Mu of Zhou on the basis of the *Xian Gui*), *Zhongguo lishi wenwu* 中國歷史文物 (*Chinese historical cultural relics*) 2006.3: 9-10.

"Texts Lost in Texts: Recovering the 'Zhai Gong' Chapter of the *Yi Zhou Shu*," in *Studies in Chinese Language and Culture: Festschrift in Honour of Christoph Harbsmeier on the Occasion of His 60th Birthday*, ed. Christoph Anderl and Halvor Eifring. Oslo: Hermes Academic Publishing, 2006. Pp. 31-47.

"The Guodian Manuscripts and their Place in Twentieth-Century Historiography on the *Laozi*," *Harvard Journal of Asiatic Studie*s 65.2 (December 2005): 417-57.

"A First Reading of the Shanghai Museum *Zhou Yi* Manuscript," *Early China* 30 (2005): 1-24.

"*Zhushu jinian* de zhengli he zhengliben" 《竹書紀年》的整理和整理本 (The editing and editions of the *Bamboo Annals*), in *Chutu wenxian yanjiu fangfa lunwenji* 出土文獻研究方法論文集 (*Essays on research methods for unearthed texts*), ed. Cai Guoliang 蔡國良, Zheng Jixiong 鄭吉雄, and Xu Fuchang 徐富昌. Taipei: Taiwan daxue chuban zhongxin, 2005. Pp. 339-441.

"Ren zhi si ye qing, zhi zhi si ye chang" 仁之思也清知之思也悵 (Aspirations of humaneness are clarifying, aspirations of wisdom are extending), *Zhongguo xueshu* 中國學術 (*Chinese scholarship*): 5.3-4 (2004): 381-89.

"Shi lun Zi yi cuo jian zhengju ji qi zai Li ji ben 'Zi yi' bianzuan guocheng de yuanyin he houguo" 試論緇衣錯簡證據及其在禮記本緇衣編纂過程的原因和後果 (A trial discussion of the evidence for misplaced strips in the Black Jacket and its reasons and consequences in the editorial process of the 'Black Jacket' chapter of the *Record of Ritual*), in *Xin chutu wenxian yu gudai wenming yanjiu* 新出土文獻與古代文明研究 (*Research on newly unearthed texts and ancient civilization*), ed. Zhu Yuanqing 朱淵清 and Xie Weiyang 謝維揚. Shanghai: Shanghai daxue chubanshe, 2004. Pp. 287-96.

"Afterword," in T.H. Tsien, *Written on Bamboo and Silk*, revised second edition (Chicago: The University of Chicago Press, 2004), pp. 207-32. Translated into Chinese by Wang Zhengyi 王正義, as "1960 nian yilai Zhongguo guwenzixue de fazhan" 1960 年以來中國古文字學的發展 (The development of Chinese paleography since 1960), *Wenxian* 文獻 (Texts): in press; *Zhonghua minguo tushuguan xuehui huibao* 中華民國圖書館學會會報 (*Bulletin of the Library Association of China*) 74 (June 2005): 51-68; *Wenxian* 文獻 (Texts) 2005.4: , 2006.1:.

"42 nian 43 nian liangge *Yu Lai ding* de niandai" 42 年 43

年兩個吳來鼎的年代 (The date of the 42nd-year and 43rd-year *Yu Lai ding*), *Zhongguo lishi wenwu* 中國歷史文物 (*China's historical cultural relics*) 2003.5: 49-52.

"Shang bo xin huo *Da Zhu Zhui ding* dui Xi Zhou duandai yanjiu de yiyi" 上博新獲大祝追鼎對西周斷代研究的意義 (The significance of the Shanghai Museum's newly acquired *Da Zhu Zhui ding* for the study of Western Zhou chronology), *Wenwu* 文物 (*Cultural relics*), 2003.5: 45-47.

"Shang wang Wu Ding de moqi: Zhongguo shang gu niandaixue de chonggou shiyan" 商王武丁的末期：中國上古年代學的重構實驗 (The last years of Shang king Wu Ding: An experiment in the reconstruction of China's ancient chronology), translated by Zhang Deshao 張德劭, *Huadong Shifan daxue Zhongguo wenzi yanjiu yu yingyong zhongxin xuebao* 華東師範大學中國文字研究與應用中心學報 (*Journal of the Center for the Study and Application of Chinese Characters, East China Normal University*) 2003: 1-27.

"The Fuyang Zhou Yi and the Making of a Divination Manual," *Asia Major*, 3rd ser. 14.1 (2001 [actually 2003]): 7-18. French translation: "Le Zhou Yi de Fu Yang et l'ebaboration d'un manuel de divination," Traduction de l'anglais : J.P. De Leeck. *Djofil* 38 (Juin 2007): 1-11.

"Hoards and Family Histories in Qishan County, the Zhouyuan, during the Western Zhou Dynasty," in *New Perspectives on China's Past: Chinese Archaeology in the Twentieth Century,* ed. Xiaoneng Yang (New Haven: Yale University Press, with the Nelson-Atkins Museum of Art, 2004), pp. 254-67.

"Toward a Social Geography of the Zhouyuan during the Western Zhou Dynasty: The Jing and Zhong Lineages of Fufeng County," in *Political Frontiers, Ethnic Boundaries, and Human Geographies in Chinese History*, eds. Nicola di Cosmo and Don J. Wyatt (London: RoutledgeCurzon, 2003), pp. 16-34.

"Jin Chu Gong ben zu kao: Jianlun *Zhushu jinian* de liangge zuanben" 晉出公奔卒考：兼論竹書紀年的兩個纂本 (A Study of Jin Chu Gong's exile and death: With a discussion of the two editions of the *Bamboo Annals*), *Shanghai bowuguan jikan* (Shanghai Museum Bulletin), 9 (2002): 186-94.

"The Wangjiatai *Gui Cang*: An Alternative to *Yi jing* Divination," in Facets of *Tibetan Religious Tradition and Contacts with Neighbouring Cultural Areas*, ed. A. Cadonna and E. Bianchi, *Orientalia Venetiana* 12 (Firenze, 2002): pp. 95-126.

"New Sources of Western Zhou History: Recent Discoveries of Inscribed Bronze Vessels," *Early China* 26-27 [2001-2002]: 73-98.

"Jin Hou de shixi ji qi dui Zhongguo gudai jinian de yiyi" 晉侯的世系及其對中國古代紀年的意義 (The Genealogy of the Lords of Jin and Its Significance for the Chronology of Ancient China), with Ni Dewei 倪德衛 (David S. Nivison), *Zhongguo shi yanjiu* 中國史研究 (*Research on Chinese History*), 2001.1: 3-10. A revised English version is: "The Jin Hou Su Bells

Inscription and Its Implications for the Chronology of *Early China*," with David S. Nivison, Early China 25 (2001): 29-48.

"Wang Bi de Huang Lao zhengzhi sixiang" 王弼的黃老政治思想 (Wang Bi's Huang-Lao Political Philosophy), in *Di er jie Dao jiao wenhua guoji yantaohui lunwenji* 第二屆道教文化國際研討會論文集(*Papers from the Second International Conference on Daoist Culture*; Hong Kong, 1999), forthcoming.

"Xici zhuan de bianzuan" 繫辭傳的編纂 (The Redaction of the *Commentary on the Appended Statements*), in *Wenhua de yizeng: Hanxue yanjiu guoji huiyi lunwenji; Zhexue juan* 文化的饋贈： 漢學研究國際會議論文集；哲學卷 (*Offerings of Culture: Essays from the International Conference on Chinese Studies*; Philosophy Issue), ed. Beijing daxue Zhongguo chuantong wenhua yanjiu zhongxin (Beijing: Beijing daxue chubanshe, 2000), pp. 262-7; republished, without authorization, as "Boshu *Xici zhuan* de bianzuan" 帛書繫辭傳的編纂 (The Redaction of the Silk Manuscript *Commentary on the Appended Statements*), *Daojia wenhua yanjiu* 道家文化研究 (*Researches on Daoist Culture*) 18 (2000): 371-81. An English version is "The Writing of the Xici Zhuan and the Making of the Yijing," in *Measuring Historical Heat: Event, Performance and Impact in China and the West, Symposium in Honour of Rudolf G. Wagner on his 60th Birthday, Heidelberg, November 3rd-4th 2001*, at http://www.sino.uni-heidelberg.de/conf/symposium2.pdf, pp.

197-221 (unfortunately, all of the Chinese characters appear garbled).

"Introduction," "The Harmony of Heaven and Earth," "The Realm of the Ghosts and Spirits," and "Afterword," in *China: The Land of the Heavenly Dragon* (London: Duncan Baird Publishers, 2000), pp. 6-11, 120-35, 136-45, 230-37.

"Bronzes from Hoard 1 at Zhuangbai, Fufeng, Shaanxi Province," in *The Golden Age of Archaeology: Celebrated Archaeological Finds from the People's Republic of China* (Washington: National Gallery of Art, 1999), 236-47.

"Fu bu fu, zi bu zi: Shilun Xi Zhou zhongqi Xun gui he Shi You gui de duandai" 父不父，子不子：試論西周中期詢簋和師酉簋的斷代 (The Father is Not the Father, The Son is Not the Son: A Trial Discussion of the Periodization of the Mid-Western Zhou Xun gui and Shi You gui), *Zhongguo guwenzi yu gu wenxian* 中國古文字與古文獻 (*Chinese Paleography and Ancient Texts*) 1 (1999): 62-64.

"Introduction" (with Michael Loewe), in *The Cambridge History of Ancient China: From the Origins of Civilizations to 221 B.C.*, eds. Michael Loewe and Edward L. Shaughnessy (New York: Cambridge University Press, 1999), pp. 1-18, 19-29, 292-351. Translated as "*Jianqiao Zhongguo gudai shi qianyan*" 劍橋中國古代史前言 (Introduction to *The Cambridge History of Ancient China*), *Zhonghua wenshi luncong* 中華文史論叢 (*Essays on China's Cultural History*) 2007: 1-20.

"Calendar and Chronology," in *The Cambridge History of Ancient China: From the Origins of Civilizations to 221 B.C.*, eds. Michael Loewe and Edward L. Shaughnessy (New York: Cambridge University Press, 1999), pp. 19-29.

"Western Zhou History," in *The Cambridge History of Ancient China: From the Origins of Civilizations to 221 B.C.*, eds. Michael Loewe and Edward L. Shaughnessy (New York: Cambridge University Press, 1999), pp. 292-351.

"Commentary, Philosophy, and Translation: Reading Wang Bi's Commentary to the *Yi jing* in a New Way," Early China 22 (1997): 221-245.

"How the Poetess Came to Burn the Royal Chamber," in *Before Confucius: Studies in the Creation of the Chinese Classics* (Albany, N.Y.: SUNY Press, 1997), pp. 221-38.

"Xi Zhou zhu wang niandai" 西周諸王年代 (The dates of Western Zhou kings), in *Xi Zhou zhu wang niandai yanjiu* 西周諸王年代研究 (*Studies of the dates of Western Zhou kings*), eds. Zhu Fenghan 朱鳳瀚 and Zhang Rongming 張榮明 (Guiyang: Guizhou Renmin chubanshe, 1998), pp. 268-292.

"Introduction" and "Western Zhou Bronze Inscriptions," in *New Sources of Early Chinese History: An Introduction to the Reading of Inscriptions and Manuscripts*, ed. Edward L. Shaughnessy (Berkeley: Institute of East Asian Studies and the Society for the Study of Early China, 1997), pp. 1-19, 57-84.

"Yan guo tongqi zu kao chenghao yu Zhou ren shifa de

qiyuan" 燕國銅器祖考稱號與周人諡法的起源 (Names of ancestors in bronzes from the state of Yan and the origin of the Zhou epithet system), in *Beijing jian cheng 3040 nian ji Yan wenming guoji xueshu yantaohui huiyi zhuanji* 北京建城 3040 年暨燕文明國際學術研討會會議轉輯 (*Special monograph from the conference International Conference on the 3040th Anniversary of the Founding of Beijing and on Yan Culture*) (Beijing: Beijing Yanshan chubanshe, 1997), pp. 320-326.

"Military Histories of *Early China*: A Review Article," Early China 21 (1996): 159-82.

"The Origin of an *Yijing* Line Statement," *Early China* 20 (1995): 223-240.

"Micro-Periodization and the Calendar of a Shang Military Campaign," in *Chinese Language, Thought, and Culture*, ed. P.J. Ivanhoe (Peru, Illinois: Open Court Press, 1996), pp. 58-82; a Chinese version published as "Yinxu buci de weixi duandaifa: Yi Wu Ding shidai de yici zhanyi wei li" 殷虛卜辭的微細斷代法：以武丁時代的一次戰役為例 (Micro-periodization of Yinxu oracle-bone inscriptions: Taking A Wu Ding-period military campaign as an example), in *Jiaguwen faxian yibai zhounian xueshu yantaohui lunwenji* 1898-1998 甲骨文發現一百周年學術研討會論文集1898-1998 (*Collected Papers of the scholarly conference on the one-hundred anniversary of the discovery of oracle-bone inscriptions*) (Taipei: Taiwan Shifan daxue Guowenxi and Zhong yan yuan Lishi yuyan yanjiusuo, 1998), pp. 31-44.

"Lüelun jinwen Shang shu Zhou shu gepian de zhuzuo niandai" 略論今文尚書周書各篇的著作年代 (An outline discussion of the dates of composition of the new text chapters in the Zhou Documents section of the *Book of Documents*), in *Dierjie Guoji Zhongguo guwenzi yantaohui lunwenji, xubian* 第二屆國際中國古文字研討會論文集續編 (*Collected Papers of the Second International Conference on Chinese Paleography*) (Hong Kong: Chinese University of Hong Kong, 1996), pp. 399-404.

"Xi Zhou zhi shuaiwei" 西周之衰微 (The decline of the Western Zhou), in *Jin xin ji: Zhang Zhenglang xiansheng bashi qingshou lunwenji* 盡心集：張政烺先生八十慶壽論文集 (*Exerting the Heart: Essays Celebrating the Eightieth Birthday of Mr. Zhang Zhenglang*), ed. Wu Rongzeng吳榮曾 (Beijing: Zhongguo Shehui kexue chubanshe, 1996), pp. 120-126.

"Shanggu lishi yanjiu de sange jiben yuanze--Cong Xi Zhou shidai Zhou wangchao yu Nan Huai Yi de guanxi tanqi"上古歷史研究的三個基本原則：從西周時代周王朝與南淮夷的關係談起 (Three basic principles of research on ancient history--based on the relations between the Zhou court and the Southern Huai Yi during the Western Zhou period), *Zhongguo shanggu Qin Han xuehui tongxun* 中國上古秦漢學會通訊 (*Bulletin of the Society for the Study of China's Antiquity and Qin and Han Dynasties*) 1 (1995): 1-8.

"From Liturgy to Literature: The Ritual Contexts of the Earliest Poems in the Classic of Poetry," *Hanxue yanjiu* 漢學研究 (*Chinese Studies*) 13.1 (1994): 133-164. A Chinese version published as "Cong Xi Zhou li zhi gaige kan Shi jing Zhou Song de yanbian" 從西周禮制改革看詩經周頌的演變, *Hebei shiyuan xuebao* 河北師院學報 1996.

"Chinese Religions: The State of the Field, Part I: Early Religious Traditions: The Neolithic Period through the Han, ca. 4000 B.C.E. to 220 C.E.: Western Chou Period," *Journal of Asian Studies* 54.1 (February 1995): 145-148.

"A First Reading of the Mawangdui *Yijing* Manuscript," *Early China* 19 (1994): 47-73.

"*I Ching*," "*I Chou shu*" and "*Shu ching*," in *Early Chinese Texts: A Bibliographic Guide*, ed. Michael Loewe (Berkeley: Institute of East Asian Studies, 1993), pp. 216-228, 229-233, 376-389.

"Zhougong ju dong xin shuo--Jianlun Shao gao Jun Shi zhuzuo beijing he yizhi" 周公居東新說——簡論召誥君奭著作背景和意旨 (A new explanation of the Duke of Zhou's residence in the east--with a discussion of the background and purpose of the composition of the 'Shao gao' and 'Jun Shi' [chapters of the *Book of Documents*]), in *Dierci Xi-Zhou shi xueshu taolunhui lunwenji* 第二次西周史學術討論會論文集 (*Essays from the Second Conference on Western Zhou History*), ed. Shaanxi Lishi bowuguan (Xi'an: Shaanxi Renmin jiaoyu chubanshe, 1993), pp. 872-887. A revised and expanded English version published as "The Duke of Zhou's Retirement in the East and the Beginnings of the Minister-Monarch

Debate in Chinese Political Philosophy," *Early China* 18 (1993): 41-72.

"Marriage, Divorce and Revolution: Reading between the Lines of the *Book of Changes*," *Journal of Asian Studies* 51.3 (August 1992): 587-599. Translated into Chinese as "Jiehun lihun yu geming--Zhou Yi de yan wai zhi yi" 結婚, 離婚與革命——周易的言外之意" (translated by Li Hengmei 李衡眉 and Guo Mingqin 郭明勤), *Zhouyi yanjiu* 周易研究 20 (1994): 45-57.

"Zhijiage daxue suocang Shangdai jiagu" 芝加哥大學所藏商代甲骨 (Shang dynasty oracle bones in the collection of the University of Chicago), in *Zhongguo tushu wenshi lunwenji* 中國圖書文史論文集 (*Collected Essays on Chinese Bibiliography, Literature and History*), ed. Ma Tai-loi 馬泰來 (Taipei: Zhengzhong shuju, 1991), pp. 197-207; (Beijing: Xiandai chubanshe, 1992), pp. 231-243.

"Jianlun Bao you de zuozhe wenti" 簡論保卣的作者問題 (A brief discussion regarding the maker of the Bao you), *Shanghai bowuguan jikan* 上海博物館館刊 5 (1990): 99-102.

"Western Zhou Civilization: A Review Article," *Early China* 15 (1990): 197-204.

"Ci ding mingwen yu Xi-Zhou wanqi niandai kao" 此鼎銘文與西周晚期年代考 (The inscription on the "*Ci ding*" and the date of the late Western Zhou), *Dalu zazhi* 大陸雜誌 80.4 (1990): 16-24; reprinted in *Xi Zhou zhu wang niandai yanjiu* 西周諸王年代研究 (Studies of the dates of Western Zhou kings), ed. Zhu Fenghan 朱鳳瀚 and Zhang Rongming 張榮明 (Guiyang:

Guizhou Renmin chubanshe, 1998), pp. 248-257.

"The Role of Grand Protector Shi in the Consolidation of the Zhou Conquest," *Ars Orientalis* 24 (1989): 51-77.

"Historical-Geography and the Extent of the Earliest Chinese Kingdoms," *Asia Major* 2.2 (November, 1989): 1-22.

"Shang Oracle-Bone Inscriptions," in *Ritual and Reverence: Chinese Art at the University of Chicago* (Chicago: Smart Gallery of Art, 1989), pp. 68-90.

"Shishi Zhouyuan buci si zi - jianlun Zhoudai zhenbu zhi xingzhi" 試釋周原卜辭 ⊕ 字——兼論周代貞卜之性質 (A preliminary interpretation of the Zhouyuan oracle-bone graph si: with a discussion of the nature of Zhou period divination), *Guwenzi yanjiu* 古文字研究 17 (1989): 304-308.

"Western Cultural Innovations in China, 1200 B.C.," *Sino-Platonic Papers* 11 (July 1989).

"Zhouyi shifa yuan wu zhigua kao" 周易筮法原無之卦考 (That the *Zhouyi* divination method originally did not have moving hexagrams), *Zhouyi yanjiu* 周易研究 1 (1988): 15-19.

"Shuo Qian zhuan zhi, Kun xi pi xiang yi" 說乾專直，坤翕闢象意 (On Qian's being curled and straight, Kun's being closed and open), *Wenshi* 文史 30 (July, 1988): 24.

"The Historical Significance of the Introduction of the Chariot into China," *Harvard Journal of Asiatic Studies* 48.1 (June, 1988): 189-237. Published in Chinese as "Zhongguo mache de qiyuan ji qi lishi yiyi" 中國馬車的起源及其歷史意義, *Hanxue yanjiu*

漢學研究 7.1 (June, 1989): 131-162.

"The 'Current' *Bamboo Annals* and the Date of the Zhou Conquest of Shang," *Early China* 11-12 (1985-87): 33-60. Published in Chinese as "Zhushu jinian yu Wu wang ke Shang de niandai" 《竹書紀年》與武王克商的年代 (The *Bamboo Annals* and the date of King Wu's conquest of Shang), *Wenshi* 文史 38 (1994): 7-18; reprinted in *Wu wang ke Shang zhi nian yanjiu* 武王克商之年研究 (*Researches on the date of King Wu's conquest of Shang*), ed. Song Jian 宋健 (Beijing: Beijing Shifan daxue chubanshe, 1997), pp. 445-453.

"Western Zhou Oracle-Bone Inscriptions: Entering the Research Stage?" *Early China* 11-12 (1985-87): 146-194.

"Cong Jufu xu gai mingwen tan Zhou wangchao yu Nan Huaiyi de guanxi" 從駒父蓋銘文談周王朝與南淮夷的關係 (Relations between the Zhou court and the Southern Huaiyi as seen from the inscription on the cover of the *Jufu xu*), *Hanxue yanjiu* 漢學研究 5.2 (December, 1987): 567-573; reprinted in *Kaogu yu wenwu* 考古與文物 1988.5: 95-98.

"Zaoqi Shang-Zhou guanxi ji qi dui Wu Ding yihou Yin-Shang wangshi shili fanwei de yiyi" 早期商周關係及其對武丁以後王室勢力範圍的意義 (Early Shang-Zhou relations and their significance for the sphere of influence of the post-Wu Ding Shang royal court), *Jiuzhou xuekan* 九州學刊 1.4 (September, 1987): 19-32. Preliminary version published without authorization in *Guwenzi yanjiu* 古文字研究 13 (June, 1986; actually 1987): 129-143.

"Zhouyi Qian gua liulong xinjie" 周易乾卦六龍新解(An explanation of the six dragons of Qian hexagram of the *Zhouyi*), *Wenshi* 文史 24 (1986): 9-14.

"On the Authenticity of the *Bamboo Annals*," *Harvard Journal of Asiatic Studies* 46.1 (June 1986): 149-180. Published in Chinese as "Ye tan Zhou Wuwang de zunian - jianlun Jinben zhushu jinian de zhenwei" 也談周武王的卒年——兼論今本《竹書紀年》的真偽, *Wenshi* 文史 29 (1988): 7-16.

"Ceding Duo You ding de niandai" 測定多友鼎的年代, *Kaogu yu wenwu* 考古與文物 1985.6: 58-60. Published in English as "The Date of the 'Duo You ding' and its Significance," *Early China* 9-10 (1984-85): 55-69.

"Shi 'yu fang'" 釋御方 (Explanation of "*yufang*" [to defend the borderlands]), *Guwenzi yanjiu* 古文字研究 9 (1984): 97-109.

"Recent Approaches to Oracle-Bone Periodization: A Review," *Early China* 8 (1983-84): 1-13.

"'New' Evidence on the Zhou Conquest," *Early China* 6 (1981-82): 55-81.

Short Book Reviews

Ralph D. Sawyer. With the bibliographic collaboration of Mei-chün Sawyer. *Conquest and Domination in Early China: Rise and Demise of the Western Chou. The Journal of Military History* 78.3 (1914): in press.

Matthias Richter. *The Embodied Text: Establishing Textual Identity in Early Chinese Manuscripts,* in *Journal of*

Chinese Studies (Institute of Chinese Studies, Chinese University of Hong Kong Kong) 60 (2014): in press.

Geoffrey Sampson. *Love Songs of Early China* (Donnington: Shaun Tyas, 2006), in *Modern Philology* 106.2 (November 2008): 197-200.

Lothar von Falkenhausen. *Chinese Society in the Age of Confucius (1000-250 BC): The Archaeological Evidence* (Los Angeles: Cotsen Institute of Archaeology, University of California, Los Angeles, 2006), in *Journal of Asian Studies* 66.4 (2007): 1129-32.

Kidder Smith, Jr., Peter K. Bol, Joseph A. Adler, and Don J. Wyatt ed.. *Sung Dynasty Uses of the* I Ching (Princeton, N.J.: Princeton University Press, 1990), in *Journal of Religion* 72.1 (1992): 146-47.

Axel Schuessler. *A Dictionary of Early Zhou Chinese* (Honolulu: University of Hawai'I Press, 1987), in *Philosophy East & West* 40.1 (1990): 110-12.

K.C. Chang ed.. *Studies of Shang Archaeology* (New Haven, Conn.: Yale University Press, 1986), in *Journal of the American Oriental Society* 107.3 (1987): 500-03.

Translations, Reports, and Occasional Pieces

Yang Hanqun 楊漢群. "Hanxue wu fen guoji guojie: Zhuanfang Meiguo Hanxuejia Xia Hanyi jiaoshou" 漢學無分國籍國界：專訪美國漢學家夏含夷教授 (In the realm of Sinology not differentiating nationality). *Guoxue xin shiye* 國學新視野 (*New perspectives on national studies*) Spring 2014: 10-18.

"Gideon's Chinese Classics," *ICS Bulletin* 2013.2. "*Jidian Zhongguo jingdian*" 吉甸中國經典 (*Gideon's Chinese classics*)，*Zhongguo Wenhua yanjiusuo tongxun* 中國文化研究所通訊 2013.2.

"Wu wu du qingtongqi mingwen" 勿誤讀青銅器銘文 (Don't misread bronze inscriptions). *Zhongguo shehui kexue bao* 中國社會科學報 (*China Social Sciences Today*) 10 August 2012. P. A-04.

Xifang Hanxuejie yu Zhongguo chutu wenxian yanjiu" 西方漢學界與中國出土文獻研究 (Western Sinology and Studies of Chinese Unearthed Texts), *Zhongguo Shehui kexue bao* 中國社會科學報 (*Chinese Social Sciences Newspaper*), 9 January 2012.

Huang Xiaofeng 黃曉峰. "Xia Hanyi tan gudai wenxian de buduan chongxie" 夏含夷談古代文獻的不斷重寫 (Edward Shaughnessy talks about ancient texts being continuously re-written). *Shanghai shuping* 上海書評 (*Shanghai Review of Books*) 2009.07.19, p. 2.

"Xuyan" 序言 (Preface), in *Yuanfang zhi shi xi*: Gudai Zhongguo *jingxuanji* 遠方之時習：古代中國精選集 (*Timely practices from distant parts: Selections from Early China*), ed. Xia Hanyi 夏含夷. Shanghai: Shanghai Guji chubanshe, 2008. Pp. 1-6.

"*Chongxie Zhongguo gudai wenxian* jielun" 重寫中國古代文獻結論 (Conclusion to *Rewriting Early Chinese Texts*), *Jianbo* 簡帛 (*Bamboo and silk*) 2 (2007): 509-14.

"The Shang Dynasty," in *World and Its Peoples: Volume 1, China and Mongolia I*. London: Brown Reference in London, in press, pp. 32-33.

"*Kongzi zhi qian* zhi hou de fansi" 《孔子之前》之後的反思 (Reflections on *Before Confucius*), *Guoji Hanxue luncong* 國際漢學論叢 (*International Papers on Sinology*), 2 (2005): 359-60.

"Ma Chengyuan 馬承源," *Early China* 29 (2004): ix-xii.

"Zhang Zhenglang 張政烺," *Early China* 29 (2004): xiii-xv. Translated by Li Ling, as "Fugao: Zhang Zhenglang (1912-2005 nian)" 訃告:張政烺(1912-2005年), in Zhang Zhenglang 張政烺, *Lun Yi cong gao* 論易叢稿 (Beijing: Zhonghua shuju, 2010), pp. 297-99.

"Zhou Dynasty," in *Das Grosse China-Lexikon*, ed. Brunhild Staiger, Stefan Friedrich and Hans-Wilm Schütte. Darmstadt: Wissenschaftliche Buchgesellschaft, 2003, pp. 885-86.

"Guwen Shang shu," "Han *Shi*," "Jinwen *Shang shu*," "Lu *Shi*," "Mao *Shi*," "Qi *Shi*," "*Shang Shu*," "*Shang shu Da zhuan*," "*Shi jing*," "Wei Guwen *Shangshu*," "*Yi jing*," and *Yi zhuan*," in *RoutledgeCurzon Encyclopaedia of Confucianism*, ed. Xinzhong Yao (London: RoutledgeCurzon, 2003), 241-2, 243, 307, 397-8, 417-8, 484, 536-7, 537-8, 552-5, 651-2, 758-61, 766-7.

"Zhi zhi buru hao zhi, hao zhi buru le zhi" 知之不如好之,好之不如樂之 (Knowing It Is Not as Good as Enjoying It, Enjoying It Is Not as Good as Delighting in It), translated by Li Ling 李零, *Zhongguo xueshu* 中國學術 (*Chinese Academics*) 1.2 (2000): 286-90.

"What Confucius Read," in Christopher Cullen, *China: The Dragon's Ascent* (London: First Media Distribution Limited, 1999), forthcoming.

"Gudai Zhongguo yanjiu chengguo, 1988-1993" 古代中國研究成果, 1988-1993, *Hanxue yanjiu tongxun* 漢學研究通訊 (*Newsletter for Research in Chinese Studies*) 51 (September 1994): 129-132; 52 (December 1994), 201-205. Also published in *Zhongguo shi yanjiu dongtai* 中國史研究動態 (*Trends in the Study of Chinese History*) 1994.11: 19-27.

"Lishi jie huo: Zhushu jinian weihe you jin gu ben zhi fen?" 歷史解惑：《竹書紀年》為何有今古本之分？ (Dispelling confusion about history: Why do we distinguish between Current and Ancient texts of the *Bamboo Annals*?), *Lishi yue ka* 歷史月刊 (*History Monthly*) 1994.3: 132-133.

"I Remember Professor Zhu Dexi," in *Zhu Dexi xiansheng jinian wenji* 朱德熙先生紀念文集 (Beijing: Yuwen chubanshe, 1993), 223-226.

"The Second Conference on Western Zhou History," *Early China News* 5 (1992): 20-22.

Translation of Qiu Xigui 裘錫圭, "Guanyu Yinxu buci de mingci shifou wenju de kaocha" 關於殷虛卜辭是否問句的考察 (*Zhongguo yuwen* 中國語文 1988.1: 1-19), *Early China* 14 (1989): 77-114.

"Wenshi dui wo de yingxiang; wo dui *Wenshi* de yinxiang" 文史對我的影響，我對文史的印象, *Shupin* 書品 1988.3: 54-57.

"1987 Shang Conference," *Early China News* 1.1: 1, 3-4, 18-19.

Translation of Liu Dajun, "A Preliminary Investigation of the Silk Manuscript Yijing," *Zhouyi Network* 1

(January 1986): 13-26.

"Brief Report on a Chinese Conference on the *Zhouyi*," *Zhouyi Network* 1 (January 1986): 4-6.

Translation of Zhu Dexi, "On the Verbality of Classical Chinese Nouns," *Early China* 9-10 (1983-85): 127-132.

Translation of Zhang Yachu and Liu Yu, "Some Observations about Milfoil Divination Based on Shang and Zhou bagua Numerical Symbols," *Early China* 7 (1981-82): 46-54.

"Fourth Annual Conference of the Chinese Paleography Association," *Early China* 7 (1981-82): 114-118.

證² ＋ 證³ ＝ 證⁵ ≡ 證 ＝ 一
（二重證據法加三重證據法等於五重證據法
當且僅當終應歸一的證據）
——再論中國古代學術證據法

著者：夏含夷
編輯：香港大學饒宗頤學術館學術部
印刷： 藝印印刷有限公司
出版：香港大學饒宗頤學術館
書號：978-988-12977-1-6
版次：2014年8月

Evidence² + Evidence³ = Evidence⁵ ≡ Evidence = One
(Double Evidence Plus Triple Evidence Equals Quintuple Evidence
If and Only If Evidence Is Unitary)
*- Further Remarks on the Evidential Method
for Scholarship on Ancient China*

Author: Edward L. Shaughnessy
Editor: Academic Division,
Jao Tsung-I Petite Ecole,
The University of Hong Kong
Printer: Standartprint Co. Ltd.
Publication: Jao Tsung-I Petite Ecole,
The University of Hong Kong
ISBN: 978-988-12977-1-6
Edition: August, 2014